增强现实游戏开发(影印版)

Augmented Reality Game Development

Micheal Lanham 著

南京　东南大学出版社

图书在版编目(CIP)数据

增强现实游戏开发:英文/(加)麦克·兰纳姆(Micheal Lanham)著.—影印本.—南京:东南大学出版社,2017.10

书名原文:Augmented Reality Game Development

ISBN 978-7-5641-7360-9

Ⅰ.①增… Ⅱ.①麦… Ⅲ.①电子计算机-游戏程序-程序设计-英文 Ⅳ.①TP317.62

中国版本图书馆 CIP 数据核字(2017)第 192683 号

图字:10-2017-122

增强现实游戏开发(影印版)

出版发行:东南大学出版社
地　　址:南京四牌楼 2 号　　邮编:210096
出 版 人:江建中
网　　址:http://www.seupress.com
电子邮件:press@seupress.com
印　　刷:常州市武进第三印刷有限公司
开　　本:787 毫米×980 毫米　　16 开本
印　　张:21
字　　数:411 千字
版　　次:2017 年 10 月第 1 版
印　　次:2017 年 10 月第 1 次印刷
书　　号:ISBN 978-7-5641-7360-9
定　　价:78.00 元

本社图书若有印装质量问题,请直接与营销部联系。电话(传真):025-83791830

Credits

Author

Micheal Lanham

Reviewer

Derek Lam

Commissioning Editor

Amarabha Banerjee

Acquisition Editor

Reshma Raman

Content Development Editor

Arun Nadar

Technical Editor

Rupali R. Shrawane

Copy Editor

Safis Editing

Project Coordinator

Ritika Manoj

Proofreader

Safis Editing

Indexer

Tejal Daruwale Soni

Graphics

Jason Monteiro

Production Coordinator

Shraddha Falebhai

About the Author

Micheal Lanham is a solutions architect with petroWEB and currently resides in Calgary, Alberta in Canada. In his current role he develops integrated GIS applications with advanced spatial search capabilities. He has worked as a professional and amateur game developer building desktop and mobile games for over 15 years. In 2007, Micheal was introduced to Unity 3D and has been an avid fan and developer ever since.

To the people I think about every day. My everything, Rhonda and children: Colton, Breann, Mikayla, and Charliegh.

About the Reviewer

Derek Lam is a game designer and Unity Certified Developer who has experience in game design on iOS and Android for over 5 years. In addition, he has a lot of experience in developing augmented reality as well as virtual reality applications. He's currently working in a construction company, producing AR and VR interactive applications, mostly for internal usage.

www.PacktPub.com

For support files and downloads related to your book, please visit www.PacktPub.com.

Did you know that Packt offers eBook versions of every book published, with PDF and ePub files available? You can upgrade to the eBook version at www.PacktPub.com and as a print book customer, you are entitled to a discount on the eBook copy. Get in touch with us at service@packtpub.com for more details.

At www.PacktPub.com, you can also read a collection of free technical articles, sign up for a range of free newsletters and receive exclusive discounts and offers on Packt books and eBooks.

https://www.packtpub.com/mapt

Get the most in-demand software skills with Mapt. Mapt gives you full access to all Packt books and video courses, as well as industry-leading tools to help you plan your personal development and advance your career.

Why subscribe?

- Fully searchable across every book published by Packt
- Copy and paste, print, and bookmark content
- On demand and accessible via a web browser

Customer Feedback

Thank you for purchasing this Packt book. We take our commitment to improving our content and products to meet your needs seriously—that's why your feedback is so valuable. Whatever your feelings about your purchase, please consider leaving a review on this book's Amazon page. Not only will this help us, more importantly it will also help others in the community to make an informed decision about the resources that they invest in to learn.

You can also review for us on a regular basis by joining our reviewers' club. **If you're interested in joining, or would like to learn more about the benefits we offer, please contact us**: `customerreviews@packtpub.com`.

Table of Contents

Preface

At the beginning of 2016, most of the world had very little knowledge of augmented reality and location-based games. That, of course, all changed with the release of Pokemon Go later that year. Literally overnight, the genre became entrenched as an upcoming trend in game development. Chances are you have played Pokemon Go and the reason you are reading this book is because of your interest in the genre of AR and location-based games.

In this book we will explore in detail the aspects of creating a location-based AR game just like Pokemon Go. Location-based AR games are expensive and require multiple services for everything from mapping to spawning monsters. However, the game we develop will be done with zero budget using freely available services. While this may not be something you could release commercially, due to some licensing restrictions, it will certainly introduce you to most of the concepts. Along the way, you will also learn how to use a great tool, Unity, and introduce many other concepts in game development.

What this book covers

`Chapter 1`, *Getting Started*, introduces the concepts that make up the genre of location-based AR games and our fictional game, Foody Go. This will be followed by a walk-through of downloading all the required software and setting up your mobile development environment with Unity.

`Chapter 2`, *Mapping the Player's Location*, starts by introducing the fundamental concepts of GIS, GPS, and mapping. Then shows how those concepts are applied to generating a real-time map and plotting the player's location in a game.

`Chapter 3`, *Making the Avatar*, builds on the previous chapter and transforms our simple location marker into a moving animated character. This allows the player to see their avatar move around the map as they move carrying their mobile device.

`Chapter 4`, *Spawning the Catch*, explains that the premise of Foody Go is about catching experimental monsters. In this chapters, we learn how to spawn the monsters around the player on the map.

`Chapter 5`, *Catching the Prey in AR*, ups the intensity by introducing integrated AR part of the game by accessing the device's camera, introduces physics for throwing balls, tracking player swiping, using creature reactions, and working with a new game scene.

`Chapter 6`, *Storing the Catch*, devoted to developing the player's inventory bag, which will hold all the Foody creatures they catch, and other useful items. Here, we walk the reader through adding persistent storage and adding a simple inventory scene.

`Chapter 7`, *Creating the AR World*, adds locations of interest around the player based on a real-time data service.

`Chapter 8`, *Interacting with an AR World*, allows the player to interact with the locations of interest. In our simple game, the player will be able to sell their caught monsters.

`Chapter 9`, *Finishing the Game*, provides the reader with the information on how to finish the game or better yet write their own location-based AR game. For the purposes of this book, we will only develop the demo Foody Go game.

`Chapter 10`, *Troubleshooting*, cover a number of troubleshooting tips and tricks to overcome those development obstacles. As with any software development exercise, problems always arise.

What you need for this book

In order to follow all the exercises in this book you will need, at a minimum, a computer capable of running Unity 5.4+ and an iOS or Android device capable of running Unity games and equipped with a GPS.

More details about Unity system requirements may be found at: `https://unity3d.com/unity/system-requirements`.

Who this book is for

This book is intended for anyone with an interest in developing their own Pokemon Go, location-based AR game. While this book assumes no prior game development skills or Unity experience, you will need a basic understanding of the C# language or equivalent (C, C++, Java, or JavaScript).

Conventions

In this book, you will find a number of text styles that distinguish between different kinds of information. Here are some examples of these styles and an explanation of their meaning.

Code words in text, database table names, folder names, filenames, file extensions, pathnames, dummy URLs, user input, and Twitter handles are shown as follows: "The next lines of code read the link and assign it to the to the `BeautifulSoup` function."

A block of code is set as follows:

```
#import packages into the project
from bs4 import BeautifulSoup
from urllib.request import urlopen
import pandas as pd
```

When we wish to draw your attention to a particular part of a code block, the relevant lines or items are set in bold:

```
<head>
<script src="d3.js" charset="utf-8"></script>
  <meta charset="utf-8">
  <meta name="viewport" content="width=device-width">
  <title>JS Bin</title>
</head>
```

Any command-line input or output is written as follows:

```
C:\Python34\Scripts> pip install -upgrade pip
C:\Python34\Scripts> pip install pandas
```

New terms and **important words** are shown in bold. Words that you see on the screen, for example, in menus or dialog boxes, appear in the text like this: "In order to download new modules, we will go to **Files** | **Settings** | **Project Name** | **Project Interpreter**."

Warnings or important notes appear in a box like this.

Tips and tricks appear like this.

Reader feedback

Feedback from our readers is always welcome. Let us know what you think about this book-what you liked or disliked. Reader feedback is important for us as it helps us develop titles that you will really get the most out of. To send us general feedback, simply e-mail `feedback@packtpub.com`, and mention the book's title in the subject of your message. If there is a topic that you have expertise in and you are interested in either writing or contributing to a book, see our author guide at `www.packtpub.com/authors`.

Customer support

Now that you are the proud owner of a Packt book, we have a number of things to help you to get the most from your purchase.

Downloading the example code

You can download the example code files for this book from your account at `http://www.packtpub.com`. If you purchased this book elsewhere, you can visit `http://www.packtpub.com/support`and register to have the files e-mailed directly to you.

You can download the code files by following these steps:

1. Log in or register to our website using your e-mail address and password.
2. Hover the mouse pointer on the **SUPPORT** tab at the top.
3. Click on **Code Downloads & Errata**.
4. Enter the name of the book in the **Search** box.
5. Select the book for which you're looking to download the code files.
6. Choose from the drop-down menu where you purchased this book from.
7. Click on **Code Download**.

Once the file is downloaded, please make sure that you unzip or extract the folder using the latest version of:

- WinRAR / 7-Zip for Windows
- Zipeg / iZip / UnRarX for Mac
- 7-Zip / PeaZip for Linux

The code bundle for the book is also hosted on GitHub at `https://github.com/PacktPublishing/ Augmented-Reality-Game-Development`. We also have other code bundles from our rich catalog of books and videos available at `https://github.com/PacktPublishing/`. Check them out!

Downloading the color images of this book

We also provide you with a PDF file that has color images of the screenshots/diagrams used in this book. The color images will help you better understand the changes in the output. You can download this file from `https://www.packtpub.com/sites/default/files/downloads/AugmentedRealityGameDevelopment_ColorImages.pdf`.

Errata

Although we have taken every care to ensure the accuracy of our content, mistakes do happen. If you find a mistake in one of our books-maybe a mistake in the text or the code-we would be grateful if you could report this to us. By doing so, you can save other readers from frustration and help us improve subsequent versions of this book. If you find any errata, please report them by visiting `http://www.packtpub.com/submit-errata`, selecting your book, clicking on the **Errata Submission Form** link, and entering the details of your errata. Once your errata are verified, your submission will be accepted and the errata will be uploaded to our website or added to any list of existing errata under the Errata section of that title.

To view the previously submitted errata, go to `https://www.packtpub.com/books/content/support`and enter the name of the book in the search field. The required information will appear under the **Errata** section.

Piracy

Piracy of copyrighted material on the Internet is an ongoing problem across all media. At Packt, we take the protection of our copyright and licenses very seriously. If you come across any illegal copies of our works in any form on the Internet, please provide us with the location address or website name immediately so that we can pursue a remedy.

Please contact us at `copyright@packtpub.com` with a link to the suspected pirated material.

We appreciate your help in protecting our authors and our ability to bring you valuable content.

Questions

If you have a problem with any aspect of this book, you can contact us at `questions@packtpub.com`, and we will do our best to address the problem.[footnote]

1
Getting Started

This chapter will introduce you to real-world adventure games—what they are, how they work, and what makes them unique. From there, we will introduce the sample real-world game that we will build through the rest of the book. Finally, after the theory, we will do a brisk walk-through on setting up a mobile development environment with Unity.

For those of you who feel that they understand the terminology of real-world adventure or augmented reality games, feel free to jump ahead to the Introducing *Foody GO* section of this chapter that will discuss the game design and concept of the sample game we will be building throughout the book.

In this chapter, we will cover the following topics:

- Defining what a real-world adventure game is
- Understanding the core elements that make a real-world adventure game
- Introducing the design of our sample game Foody GO
- Installing Unity
- Setting up Unity for mobile development
- Creating the game project

Real-world adventure games

Real-world adventure games are a genre of games that has surged in popularity recently with the release of Pokemon GO. Chances are, at the time of reading this book, you may have certainly heard of, and likely have played, the popular game. Although many think this genre is an overnight sensation, it has in fact been around for several years. Niantic, the developer of Pokemon GO, released Ingress, its first real-world game, in November 2012. The title was, and is, popular, but has only attracted a niche following of gamers, which was likely more the result of the game's complex theme rather than being specific to the genre.

Now, many can suggest that the primary catalyst that launched Pokemon GO into a gaming sensation is the Pokemon franchise combined with a new augmented reality gaming platform. Certainly, without its integral real-world interaction, Pokemon GO would have been just another popular mobile game.

So, what are the elements that make a real-world adventure or location-based augmented reality game unique?

- **Location-based**: Players have the ability to interact with virtual objects or places around them using a map. As the player physically moves in the real world, their device's GPS will update the player's location in the game, thus allowing the player to move to virtual locations and search for or interact with virtual objects or things. We will discover how to integrate the device's GPS and display a map in `Chapter 2`, *Mapping the Player's Location*.
- **Augmented Reality (AR)**: A player interacts with the real world through their device's camera. This allows them to view and interact with the virtual place or thing against the backdrop of the real world around them. Using the device's camera as a game background in order to enhance the user experience will be introduced in `Chapter 5`, *Catching the Prey in AR*.
- **Adventure game**: Players typically assume the role of an avatar driven by a mission of exploration and puzzle solving in order to ultimately reach some story-driven goal. Of course, other notable games in the real-world genre may loosely fit that definition. For the purposes of this book, we will adhere to that loose definition of adventure game. The *Introducing Foody GO* section in this chapter will cover the game design and concept of the real-world adventure we will be building through the rest of the book.

Of course, there are many other elements that will be needed to create a successful game, but essentially, location-based and augmented reality are the elements that identify the real-world adventure genre. Astute readers may notice that massively multiplayer network gameplay or MMO was omitted. Although MMO gameplay may be essential to certain game designs, it is not a requirement of this genre.

Location-based

Tracking a player's location in the real world and then overlaying that into the game's virtual world creates a unique level of immersion for players. In fact, in many real-world adventure games, warning messages are presented to players before they start playing. There are many stories of players becoming so immersed that they have hurt themselves due to an avoidable accident while playing a real-world game.

Mapping the real world on top of the game's virtual world provides new challenges to traditional mobile gaming. Developing a map interface and populating it with virtual items require some advanced GIS skills. Many developers will be new or relatively inexperienced in the concepts of GPS and GIS or rendering a map in Unity. Since mapping is a core concept to the real-world genre, it will be the basis for much of the sample game we will be building. We will spend several chapters related to the topic of maps. The following is a list of chapters that will touch on mapping and location:

- `Chapter 2`, *Mapping the Player's Location*, starts with a basic discussion of GPS and GIS and then show you how to load a map texture into a 3D Unity scene
- `Chapter 3`, *Making the Avatar*, introduces the player character avatar and shows how thc mobile's devices and the player's movement will control the avatar
- `Chapter 4`, *Spawning the Catch*, is where we start to introduce virtual items onto the map and allow the player to find those items
- `Chapter 7`, *Creating the AR World*, focuses on populating the virtual world around the player based on real-world locations
- `Chapter 8`, *Interacting with an AR World*, allows the player to interact with those virtual locations

Augmented Reality

AR has been around since 1990. The term typically covers a broad range of technologies from virtual surgical devices, Microsoft HoloLens, and mobile apps such as Snapchat. AR technologies have been slow to become mainstream in gaming until just very recently. With the advance of new technologies and the real-world adventure genre being major contributors to increasing popularity of AR in gaming.

As mentioned, AR covers a broad range of technologies or devices that provide an overlaid virtual environment to the user. However, on a mobile device, the AR experience is often a result of rendering a virtual environment over a backdrop of the device's camera. In some cases, the AR game or application will have sophisticated image processing algorithms that identity features. Those identified features may then be annotated virtually with other graphics or game options. Pokemon GO, for instance, limits the use of AR to just the camera background, whereas Snapchat provides a dynamic AR experience to the user through the use of image processing. Yet, both the game and application benefit by providing the user with a more enjoyable experience through AR.

For the purpose of the real-world adventure genre and this book, we will take a basic approach to provide the user with a basic AR experience. That means, we will look at integrating the mobile device's camera as a game backdrop. The gaming experience will be familiar to other popular games in the genre. Even with this basic approach to AR, we will still cover a number of other details and tips in the course of a few chapters. The following is a short description of the AR elements that we will work on in the subsequent chapters:

- `Chapter 5`, *Catching the Prey in AR*, will introduce using the mobile device's camera as our game's background
- `Chapter 9`, *Finishing the Game*, will discuss some ideas for ways to enhance the AR gameplay
- `Chapter 10`, *Troubleshooting*, will help you just in case things don't work as expected; this chapter will identify potential issues and provide tips to resolve them

Adventure games

Adventure games are typically characterized by a quest-driven story where players must explore and solve puzzles in order to complete the game, whereas the current real-world adventure games are more about exploration than puzzle solving and completing a quest. If anything, the current batch of real-world games are more like a **Role-Playing Game (RPG)** than a classic adventure game. In the future, we may certainly see more true classic adventure games or possibly other real-world mixes, such as real-time strategy, shooters, simulation, educational, sports, and puzzle.

In order to demonstrate how all these new concepts come together, we will be building a sample game throughout the book. This game will loosely follow an adventure style not unlike other popular titles in the real-world genre. In this game, we will be adding many common game elements, such as an avatar, character inventory, particle effects, and more in the course of several chapters; the following is a short introduction to those chapters:

- `Chapter 3`, *Making the Avatar*, helps you to add a 3D rigged and animated character to our map
- `Chapter 4`, *Spawning the Catch*, covers many concepts on GIS and mapping and features a short section on object animation
- `Chapter 5`, *Catching the Prey in AR*, introduces AR and many other gaming concepts, such as texturing, rigid body physics, player input, AI, GUI menus, and particle effects
- `Chapter 6`, *Storing the Catch*, introduces you to developing a persistent character inventory on a mobile device and more GUI development
- `Chapter 8`, *Interacting with an AR World*, helps you add additional GUI elements and more particle effects and introduces visual effect shaders
- `Chapter 9`, *Finishing the Game*, discusses the possibilities of enhancing the sample game or other ideas for a real-world game

Introducing Foody GO

Certainly, the best way to learn any new or advanced concepts is by example. **Foody GO** will be our example real-world adventure game that we will be building throughout the book. The game will follow a food theme where the player will search for and catch experimental cooking monsters. Once caught, the player must take their monsters to local restaurants in order to sell them for items, power and prestige.

Of course, our sample game will focus on the location-based augmented reality elements, but we will introduce several other technical features, as follows:

- Player mapping
- Augmented reality with the camera
- Rigged and animated 3D avatar
- Animated objects
- Simple AI
- Particle effects

- GUI menus and controls
- Persistent database storage
- Visual shader effects

We won't get into extensive detail on any of the preceding features, as most of these items could cover a book by itself. However, it will be helpful to understand how each of these elements comes together to make a real-world adventure game.

Source code

All of the books source code can be downloaded from SITE. The source code will be broken down chapter by chapter and provided as a progressive project. For each chapter, the starting and ending project's states will be provided. This will allow the more advanced reader to jump ahead in the book. Novice readers are encouraged that you follow all the examples in the book, as later chapters will be more advanced in content.

Getting into mobile development with Unity

Now, with all the background set and our course laid out, let's get started by introducing mobile development with Unity. More advanced readers who have developed Unity games on Android or iOS may want to skip ahead to `Chapter 2`, *Mapping the Player's Location*.

This installation guide is intended to be cross-platform compatible and should work on Windows, Linux, or Mac. For the sake of brevity, only screenshots for the Windows platform will be shown.

Downloading and installing Unity

Even if you have installed Unity before, but have not done mobile development, make sure that you follow this section closely. There are a couple of important steps you would not want to miss.

Perform the following steps to install Unity:

1. Open any browser and go to `https://unity3d.com/`.
2. Browse through the site and download the Unity installer for the latest stable release. If you have never downloaded Unity before, you will need to create a new account.
3. Run the Unity installer, click on **Next** to accept the license agreement and click on **Next** again.
4. On the architecture dialog, shown in the following screenshot, make sure that you select **64 bit**:

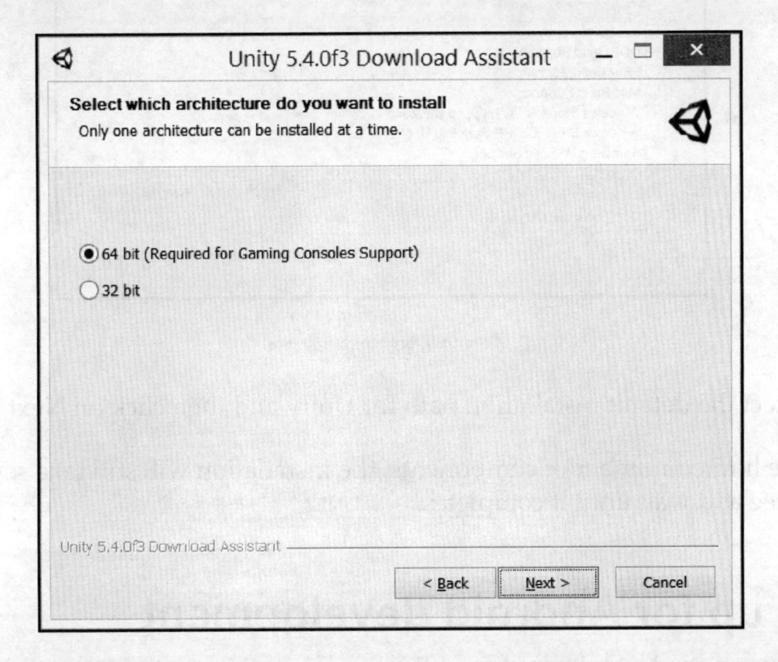

5. On the select components dialog, make sure that you choose your preferred mobile platform, Android or iOS. Many users will just select all features and install. However, it is better to be selective and install only what you need. Installing all Unity features will require about 14 GB of space, and this can quickly add up if you have installed more than one version.

In the following example screenshot, we have selected both Android and iOS. Ensure that you only select the platform you need:

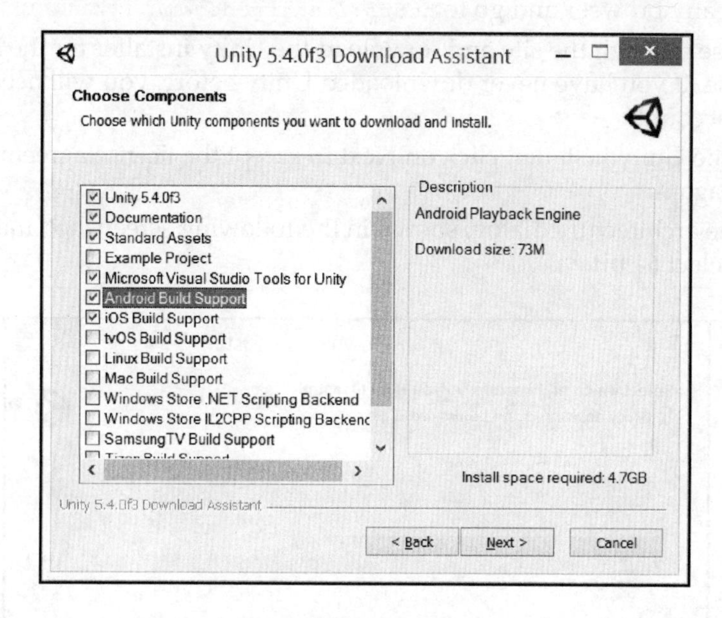

Choose only the components you need

6. Select the default installation path for Unity and then click on **Next** to install.

Even with the limited number of components, the installation will still take several minutes, so grab a coffee and wait until it completes.

Setting up for Android development

If you are using an Android device to test the game, use this section to get ready. Developers who already have experience with Android can briefly review this section or jump ahead to the *Getting Started with Unity* section in this chapter.

Always install the same architecture version, 64 or 32 bit.

Installing the Android SDK

Follow the subsequent steps to install the Android SDK on your development computer. Even if you already have the SDK installed, please review these steps to ensure that you have the right path and components set:

1. If you have not already done so, download and install the **Java Development Kit** (JDK) from
   `http://www.oracle.com/technetwork/java/javase/downloads/index.html`.

Always note where you install a development kit like the JDK or SDK.

2. Download the latest version of Android Studio from
   `https://developer.android.com/studio/index.html`.
3. After Android Studio has finished downloading, follow the instructions on `https://developer.android.com/studio/index.html` to start installation.
4. As you install Android Studio, make sure that you also install the Android SDK, as follows:

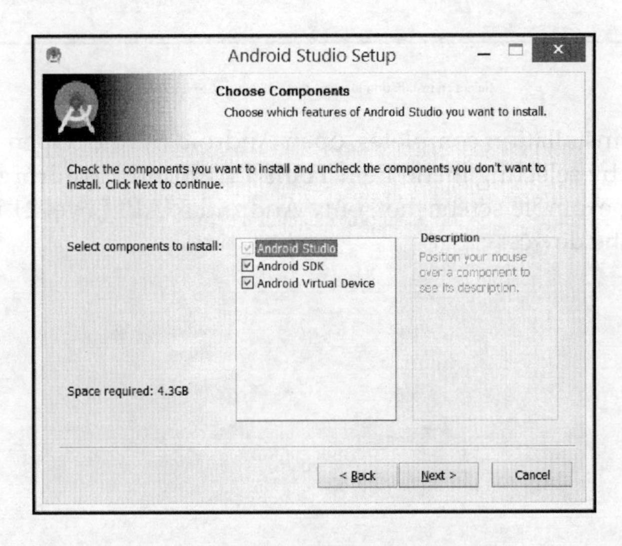

Install the Android SDK component

5. For the installation location, change the path to something that is easy for you to remember and locate. In the example screenshot, `Android/AndroidStudio` and `Android/AndroidSDK` are used:

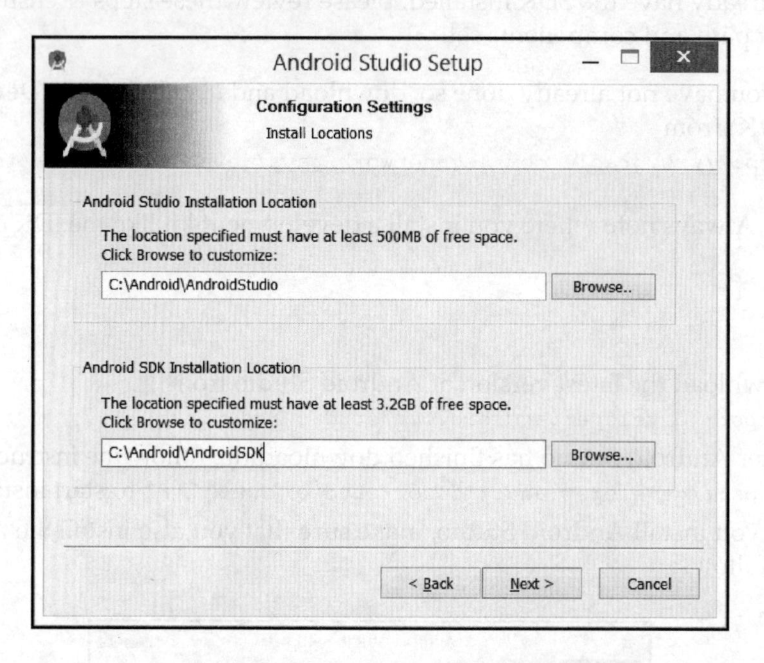

Select an installation location that will be easy to find later on

6. After the installation completes, open Android Studio. Open the **Android SDK Manager** by selecting menu item **Tools** | **Android** | **AndroidSDK**. In the following example screenshot, only Android 5, API Level 21 is selected because it matches the device:

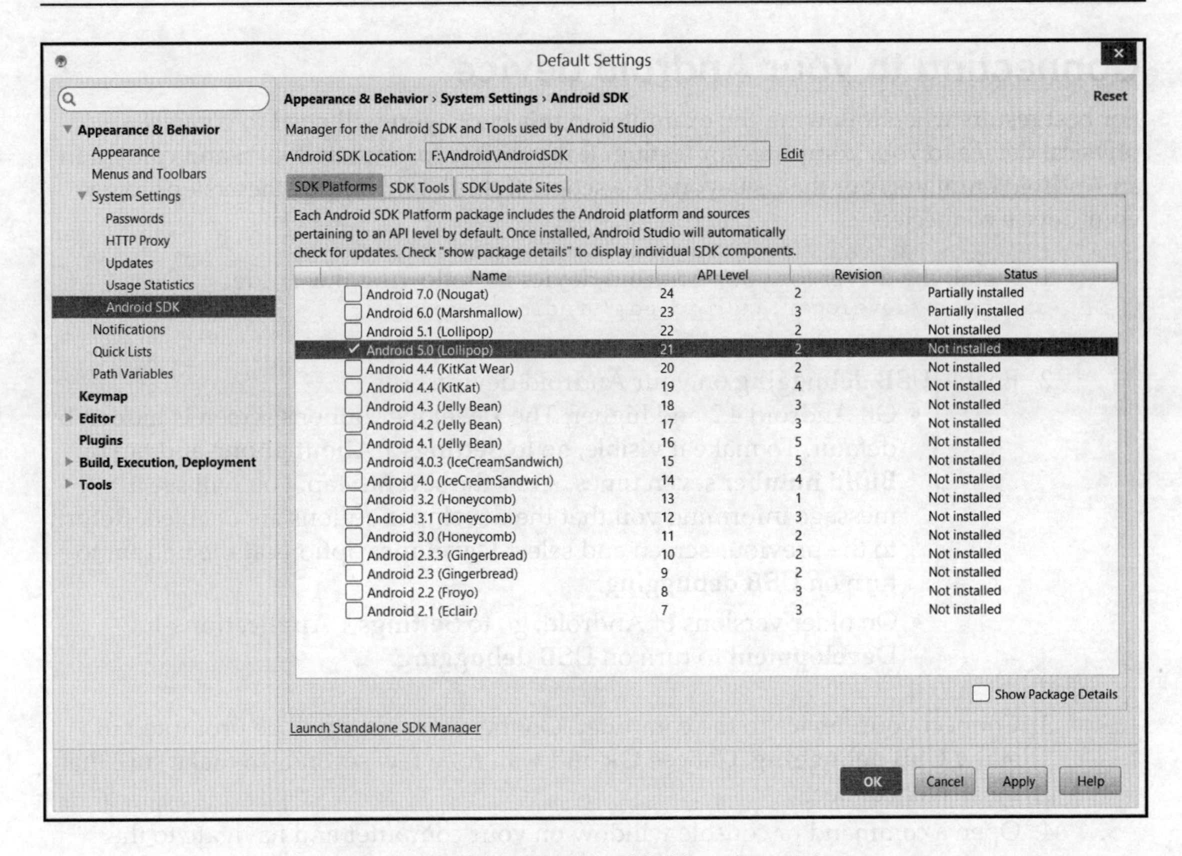

Set the Android SDK location and select API level that matches your device

7. On the **Android SDK** panel, set the location path to the same as you used in step 5. Then, select to install Android API level that matches your Android device and click on **Apply**. Locate the Android version of your phone from **Settings** | **About phone** | **Android version**. This API installation may take several minutes, so this will be a good time for another coffee or the beverage of choice.
8. After the API installation completes, close Android Studio.

Connecting to your Android device

For best results when following the examples in this book, you will need to connect a physical device to your computer for testing. It is possible to emulate a GPS and camera in an Android emulator, but that is beyond the scope of this book. Follow these steps to get your device connected:

1. Install the driver for your Android device by following this guide: `https://developer.android.com/studio/run/oem-usb.html#InstallingDriver`.
2. Enable **USB debugging** on your Android device:
    - On Android 4.2 and higher: The Developer options screen is hidden by default. To make it visible, go to **Settings** | **About phone** and tap on **Build number** seven times. After the seventh tap, you will see a message informing you that the developer options are enabled. Return to the previous screen and select **Developer** options at the bottom to turn on **USB debugging**.
    - On older versions of Android, go to **Settings** | **Applications** | **Development** to turn on **USB debugging**.
3. Connect your device to the computer. On the device you will be prompted to allow **USB debugging**. Choose **Ok** and wait for a few seconds to make sure that the driver connects.
4. Open a command or console window on your computer and navigate to the `Android/AndroidSDK` folder where we installed the **Android SDK** to above.
5. Run the following command-line commands:

```
cd platform-tools
adb devices
```

6. Your device should show up in the list. If for some reason you do not see your device in the list, consult `Chapter 10`, *Troubleshooting*. The following console window shows the commands run and sample output:

```
F:\Android\AndroidSDK\platform-tools>cd ..

F:\Android\AndroidSDK>cd platform-tools

F:\Android\AndroidSDK\platform-tools>adb devices
List of devices attached
BH90B2U116      device
```

That completes the bulk of setting up an Android device. We still have a couple more settings to make in Unity, but we will cover that in the next section on project setup.

Setting up for iOS development

In order to keep the content of this book focused and the development platform independent, we will not provide a step-by-step guide here. However, there is an excellent guide for iOS setup on the Unity site at `https://unity3d.com/learn/tutorials/topics/mobile-touch/building-your-unity-game-ios-device-testing`.

After you complete your iOS setup, return to the book, and we will begin building the example game project.

Getting started with Unity

Unity is a great platform to start learning game development or even shipping a commercial game. It is the choice of game engine for many of the popular games in the Android or iOS app stores. So, what makes Unity such a great platform to develop games on? The following is a short list of things that make Unity such a compelling platform to develop games on:

- **It is free to get started**: There is a ton of free assets or code you can use to build games. We will look at a number of free assets in this book.
- **Incredibly easy to use**: You could likely build a complete game in Unity without writing even one line of code. Fortunately, you will learn some scripting and how to write code in this book.
- **Entirely cross platform**: With Unity, you can develop on or for any environment that suits you. Of course, certain platforms, such as mobile, still have limitations and we will cover that as well.
- **Outstanding community**: Unity has a hardcore base of developers who are eager to share experiences and help others. We will be sure to showcase many of the great community resources.
- **Asset Store**: The Unity Asset Store can be an invaluable tool to build your first game or even your seventh commercial title. We will show you the deals and what to avoid.

Creating the game project

Let's get started by creating our sample game project Foody GO; we will also take this opportunity to build and deploy this starting project to your mobile device:

1. Launch Unity and start a new project named `FoodyGO`. Make sure that 3D is enabled and disable Unity analytics. Of course, you will want to save your project in an easy-to-find folder, such as Games, just as shown in the following sample screenshot:

2. Click on the **Create project** button and wait for Unity to open.
3. In the **Hierarchy** window (top-left corner), you will see a scene named **Untitled**. Beneath that, there will be the **Main Camera** and **Directional Light** attached. Here is the screenshot of what you will see:

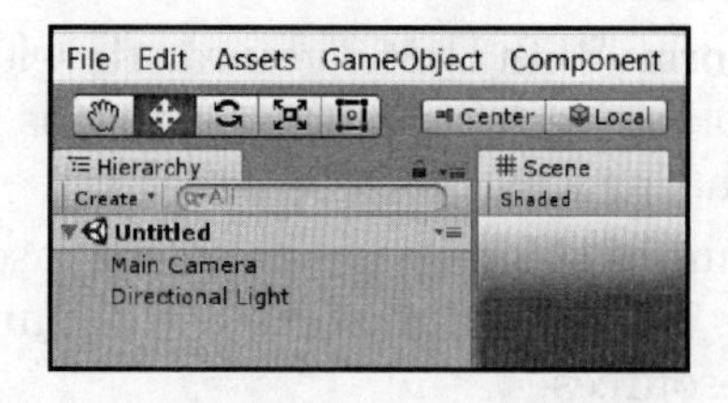

4. First thing you will want to do is rename the scene and save it. So, go to **File | Save Scene As...**
5. A save dialog will open, allowing you to choose where to save the scene. Just choose the default `Assets` folder and name your scene `Splash`. Then, click on **Save**.

6. The title of the scene should now read **Splash**. Notice that there is also a new **Splash** scene object in the project `Assets` folder.
7. Now let's adjust the Unity editor layout to match how our mobile game will run. From the menu, select **Window** | **Layouts** | **Tall**. Then, undock the **Game** tab from the main window by selecting and clicking the mouse while dragging the tab over. Then, resize the window so both the scene and game windows are roughly of the same width.
8. Save the layout by opening the menu to **Window** | **Layouts** | **Save Layout**. Name your layout `Tall_SidebySide` and click on **Save**. This will allow you to quickly return to this layout later.
9. Select the **Main Camera** by double-clicking on it, in either the **Hierarchy** window or **Scene** window. Notice how the **Scene** window will focus on the **Main Camera** object, and the **Inspector** window will display all the properties. This is what the Unity editor window should look like now:

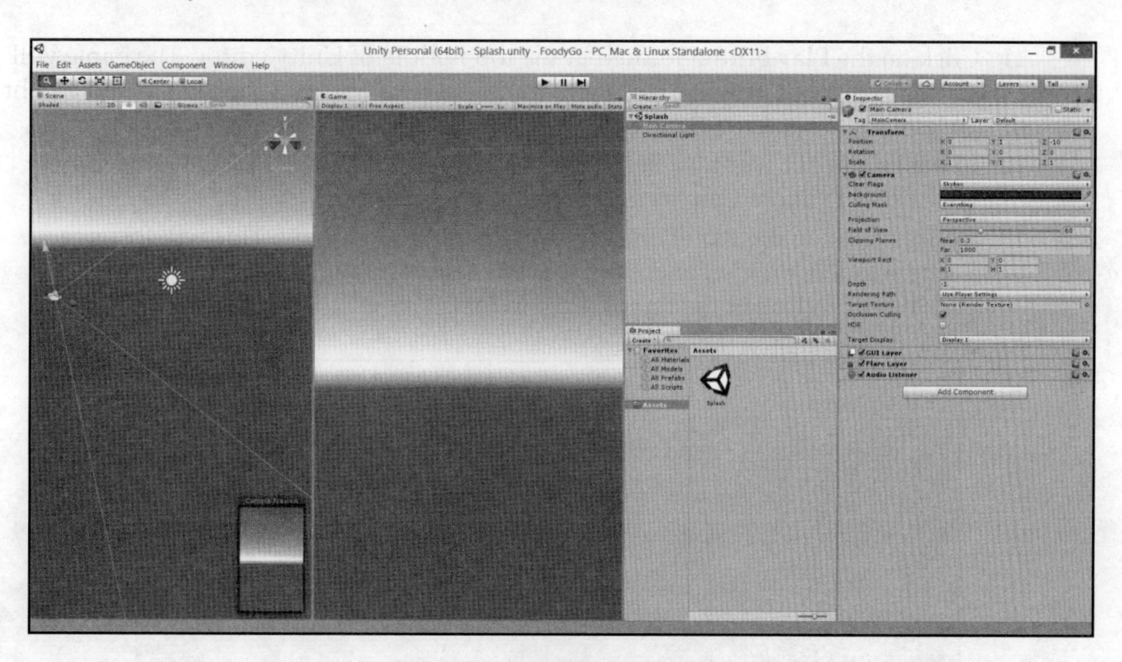

Editor layout for mobile development

10. Before we go further, let's take a look at each of the main windows we will be working with in Unity:
    - **Scene window**: This window will allow you to view and interact with the game objects in the scene.
    - **Game window**: This is the players' view of the scene rendered by the main camera.
    - **Hierarchy window**: This shows a tree view of the game objects in the scene or scenes. In most cases, you will select or add items to the scene in this window.
    - **Project window**: It provides a view of and a quick way to access the assets in your project. Not much is in our project now, but we will quickly add some new assets in the coming chapters.
    - **Inspector window**: This view allows you to inspect and alter settings on game objects.
11. Click on the **Play** button located at the top middle of Unity editor. The game will start playing, but nothing will happen because we only have a camera and a light. So let's add a simple **Splash** screen.
12. In the **Hierarchy** window, select the scene. From the menu, select **Game Object** | **UI** | **Panel** to add a **Canvas** and **Panel** to the scene.
13. Double-click on the **Panel** object in the **Hierarchy** window. This will focus the **Scene** window on the panel. In the **Scene** window, switch the view to **2D** by clicking the button at the top of the window. This is what you should see now:

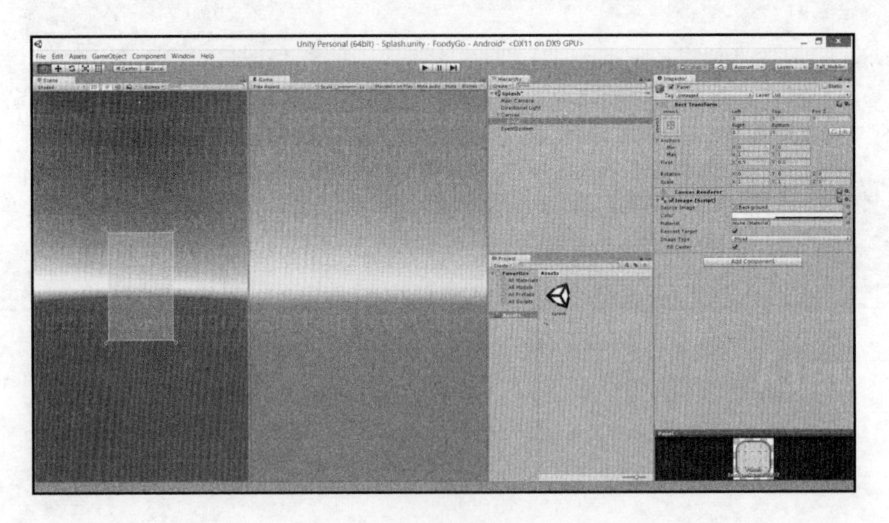

Editor window with UI focused on in Scene window

14. The panel in the scene window is a 2D UI element that allows us to render text or other content to the player. By default, the panel centers itself to the main view of the camera when it is added to the scene. This is why we see the semitransparent panel cover the entire **Game** window. Since we don't want a transparent background to the splash screen, let's change the color.
15. Select the **Panel** in the **Hierarchy** window. Then, click on the white box beside the **Color** property in the **Inspector** window to open the **Color** settings. You will see the following dialog:

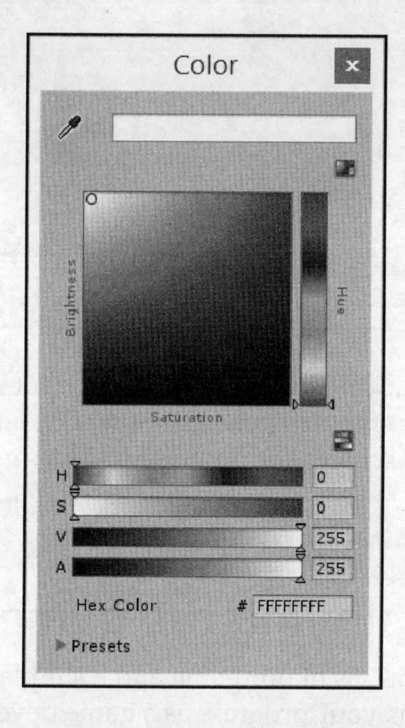

16. Enter `FFFFFFFF` in the **Hex Color** field and then close the dialog. Notice how the background of the **Game** window is now opaque white.

17. From the menu, select **Game Object | UI | Text**. In the **Inspector** window, set the properties of the **Text** to match the screen extract, as follows:

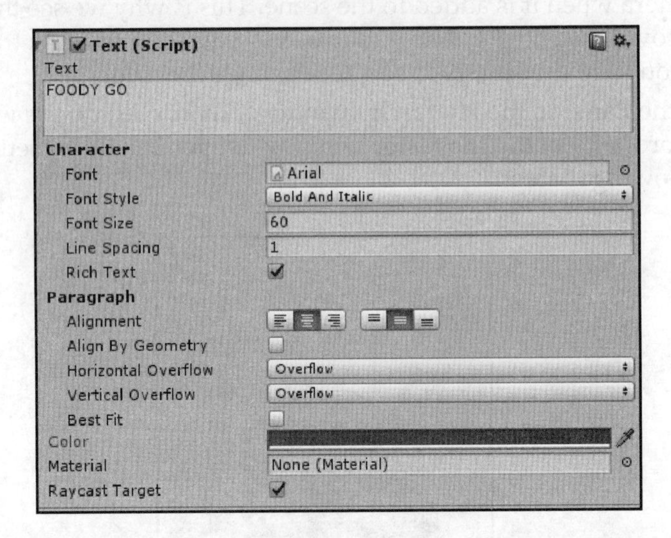

18. Run the game by clicking on the **Play** button. Not much happens, but we now have a splash screen for our game. Don't worry, later in the book, we will add some pizazz to this screen. For now though, let's get this game deployed to your device.

Building and deploying the game

Now that we have the foundations of our game and a simple splash screen let's deploy it to your device. Nothing confirms your progress as a game developer other than seeing your game run outside the Unity editor. Follow the section that is specific to your device to complete the build and deployment.

Building and deploying to Android

As long as you have followed the steps for installation in the previous sections of this chapter, deploying your game to Android should be simple. If you do encounter any deployment issues, refer to `Chapter 10`, *Troubleshooting*. Follow the next steps to build and deploy to your Android device:

1. From the menu, select **Edit | Preferences**. This will open the preferences dialog.
2. Select the **External Tools** tab. Change or set the location of your `Android SDK Path` and `Java JDK Path` to the installation paths we made a note of when installing. Close the dialog when you are finished. The subsequent sample screenshot shows where you need to enter those paths:

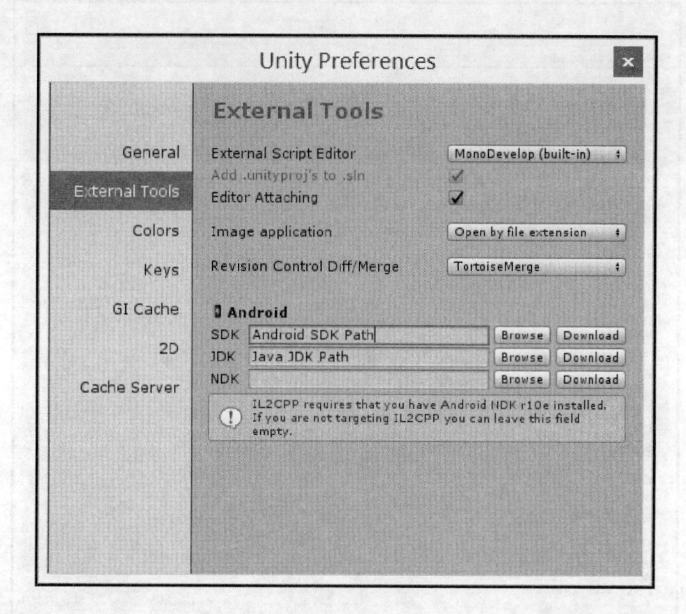

3. Select **Edit | Project Settings | Player** from the menu. Select the **Android Settings** tab and then click on **Other Settings** at the bottom of the panel. Set the **Bundle Identifier** to `com.packt.FoodyGO` as shown in the following screenshot:

4. Make sure that your Android device is connected by the USB. Refer to the *Connecting to your Android Device* section earlier in this chapter if you are unsure.

5. Open the **Build Settings** by selecting from the menu **File | Build Settings**. On the **Build Settings** dialog, click on the **Add Open Scenes** button to add the **Splash** scene. Ensure that you select the Android build from the list of build types. When you are all set and ready, click on the **Build and Run** button. This is how the **Build Settings** dialog should look like:

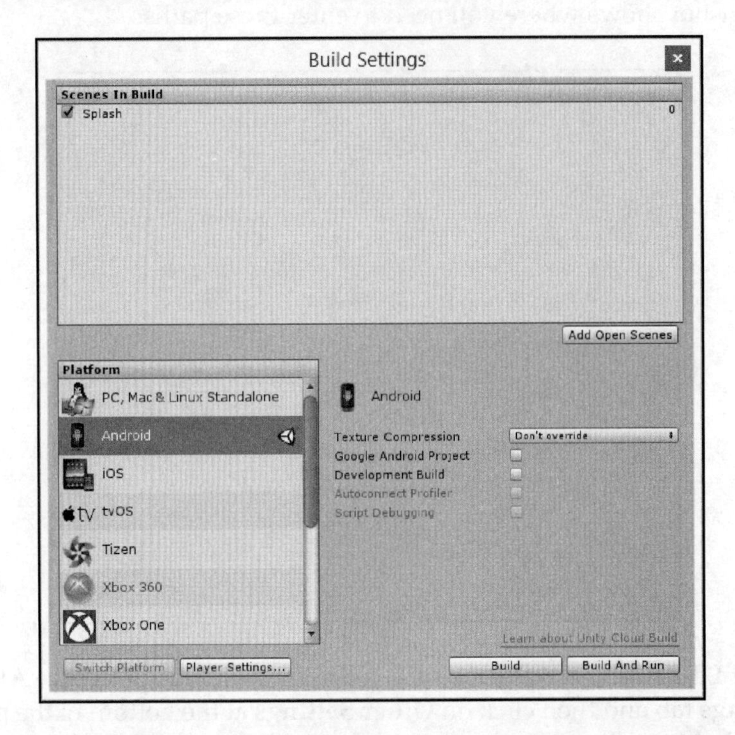

7. A file save dialog will open the project's root folder. Create a folder called `Build` while you still have the dialog box opened. Open the new **Build** folder and save your build as `com.packt.FoodyGO`. The name should match the name we used for the build identifier earlier. Then, click on **Save** to start the build.
8. Since this is the first build of the project, Unity will reimport all the project assets and other modules, which may take a few minutes. Subsequent builds should take less time, but if you change platform outputs, everything will need to be reimported.
9. After the build completes, open your device. You will likely see the Unity screen loading and then the splash screen. Congratulations, you have deployed the game to your device.

Building and deploying to iOS

Assuming you followed the previous section, *Setting up for iOS development*, you should be able to already build and deploy the game to your device. Just follow the build and deployment steps again in that page to deploy the game to your device. Download the sample project for the chapter and deploy that to your iOS device.

Summary

For this chapter, we started out by introducing what the genre of real-world adventure games is and why it has become so popular. We then went into more detail on the major components that comprise the genre and how we will be covering each component in the book. After that, we introduced the sample game Foody GO that we will be building as a sample real-world adventure game. Then, we dove headfirst into installing Unity and the required dependencies for building, deploying, and testing it on your mobile device. Finally, we created the Foody GO game project and added a simple splash screen.

In the next chapter, we will continue building on the Foody GO game project and start to add mapping. However, before we add maps to our game, we will cover some basics on GPS and GIS.

2

Mapping the Player's Location

The core of most real-world games is the location-based map. That integration of the player's real-world position mapped into the virtual world extends the game's virtual elements into the real world, allowing players to explore the world around them with a new perspective and sense of exploration.

In this chapter, we will take our first steps to understanding how to integrate maps into a Unity game. However, before working with Unity, we will cover the basics of GIS and GPS. This will allow us to establish some simple definitions and background for what can be complex topics. After that, we will dive back into Unity and add a location-based map, a basic character, and free look camera to our Foody GO project. Even though we are dealing with advanced concepts, we will keep things simple for now and avoid getting into any code. Of course, more advanced readers with a background in GIS have the options to explore the code at their leisure.

Here is a quick breakdown of what we will cover in this chapter:

- GIS terminology and basic fundamentals
- GPS terminology and fundamentals
- Google Maps
- Importing assets
- Setting up services
- Debugging with CUDLR

GIS fundamentals

GIS stands for geographic information system, a system in which geographic data is collected, stored, analyzed, manipulated, and served as maps. Although that definition is still prominent, GIS has come to mean everything from software applications, hardware, tools, science, and services. Google Maps, for example, is the best-known example of a GIS in use today. Yet, in this book, we will also use GIS to mean the science and process of converting geographic data and mapping.

Mapping

Cartographers have been mapping the world around them for thousands of years. Yet only very recently with the development of computers did it become efficient to be used in computers and GIS to create dynamic maps. Unlike traditional hand-drawn paper maps, dynamically rendered maps are composed of layers of spatial data that could describe roads, points of interest, parks, boundaries, landscape, water, and more. As much as the common Google Maps user has little control over what layers are shown, they are there. The following is a figure that shows the typical layers that could comprise a road map:

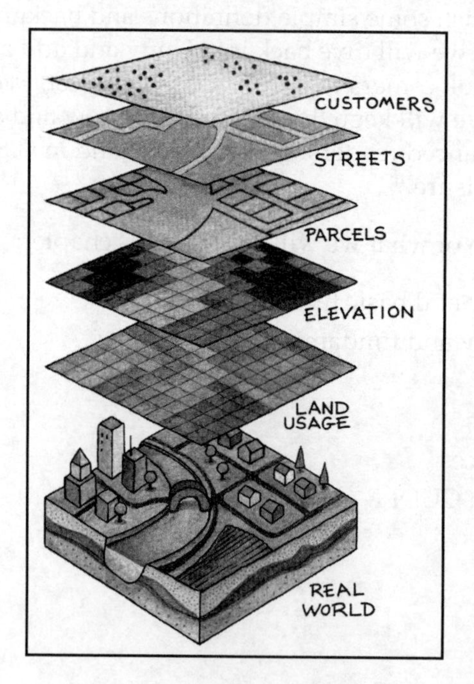

Source: `http://bit.ly/2iri2vr`

Google Maps, Bing, and other GIS providers will typically render maps at various zoom levels and then cut those maps into static image tiles. The GIS server then serves those tiles to the user. This works well for performance, but provides few options for customization or styling, aside from adding shapes, lines, or points of interest. This form of tile mapping is often referred to as static mapping. For this book, we will work with the Google Maps API, which provides us with dynamic maps.

Dynamic maps give the presenter or developer options to style and/or symbolize the data as required. For instance, on some maps, you may want to alter parks to be shown as blue instead of green. This is the type of flexibility that dynamic maps provides. When we will add the Google Maps API to our project later in this chapter, we will explore custom style and symbolization options.

Now that we understand the basics of GIS mapping, let's get into understanding some terminologies and concepts. The following is a list of terms we use when describing or making maps:

- **Map scale**: Geographical maps can represent everything from your neighborhood to the whole world. A map scale will often be rendered as text or graphic and provide the user with an understanding of the scope the map covers.
- **Zoom level**: This shares an inverse relationship with the map scale. Zoom level starts at 1, which represents a global view of the world, whereas zoom level 17 would represent a map showing your neighborhood. For our game, we will use a small map scale to allow the player to easily identify landmarks around them, which will equate to a zoom level of 17 or 18.
- **Coordinate system**: Various coordinate systems have been used across history to geographically locate points of interest. Many people will be familiar with latitude and longitude coordinates, but may not realize there are many different coordinate systems. In fact, the common latitude and longitude that we think is standard also has many variations. Since we will work with Google Maps in this book, we will use WGS 84 coordinates for everything we do. Be aware that if you do try to import data from other GIS, there could be differences that you may need to convert.

The following is a diagram showing WGS 84 major lines of interest:

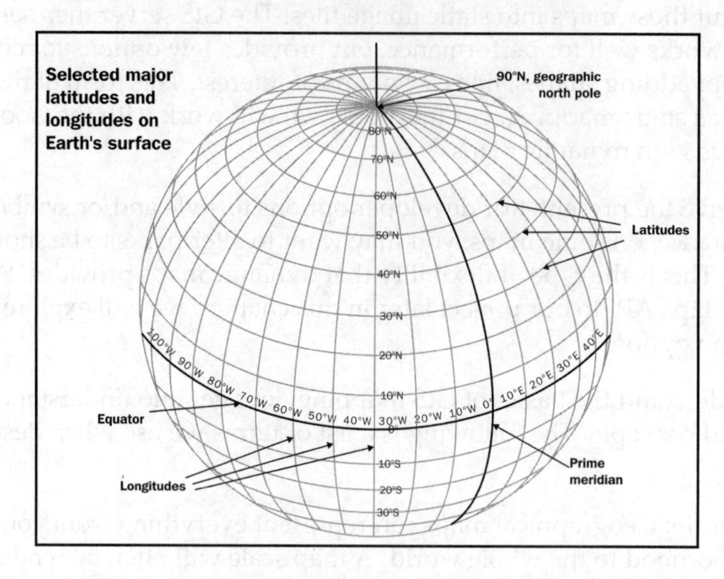

Source: `http://bit.ly/2iOoPNG`

Longitude west of the prime meridian is measured as negative. Latitude south of the equator is measured as negative.

- **Map projection**: One of the fundamental problems with mapping is representing our 3D sphere-shaped world onto a 2D service. Early cartographers resolved this issue by projecting a light from a globe onto a cylinder of paper and tracing the outline. The paper was then unrolled and a 2D view of the globe would be seen. We still use a similar method to render a view of our world. Although this 2D view is distorted as you near the poles, it has become the standard for most of our mapping needs. Suffice it to say there are many variations of map projections that are used for a variety of reasons. Other arguably better projection methods have been developed since, such as the Gall-Peters, which better accounts for the distortion at the poles. For us though, we will stick to the Google Maps' standard Google Web Mercator or just Web Mercator. The following is an example of a global map using Web Mercator projection on the left versus the Gall-Peters projection on the right:

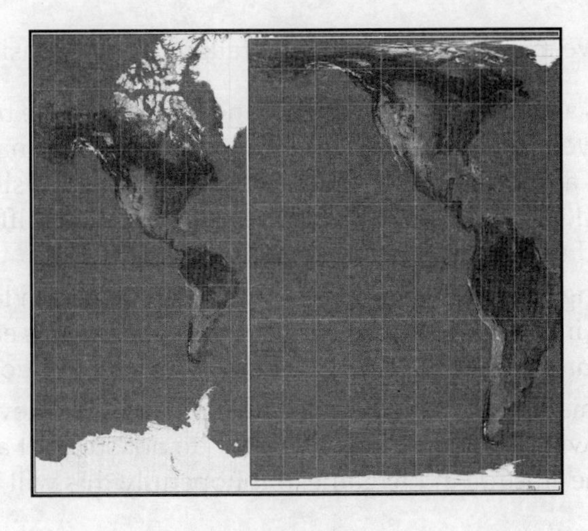

Web Mercator projection on the left and Gall-Peters on the right

GPS fundamentals

GPS is an acronym for global positioning system, a network of 24 to 32 satellites that orbit the earth twice every hour. As these satellites orbit, they emit time-encoded geographical signals as a beacon to any GPS device on earth that can see them. A GPS device then uses these signals to triangulate its position anywhere on the planet. The more satellites and signals a device can receive a signal from, the more accurate the location will be. We will get into more details about GPS triangulation and accuracy in `Chapter 4`, *Spawning the Catch.*

The following diagram shows a GPS device acquiring a signal from visible satellites on the network:

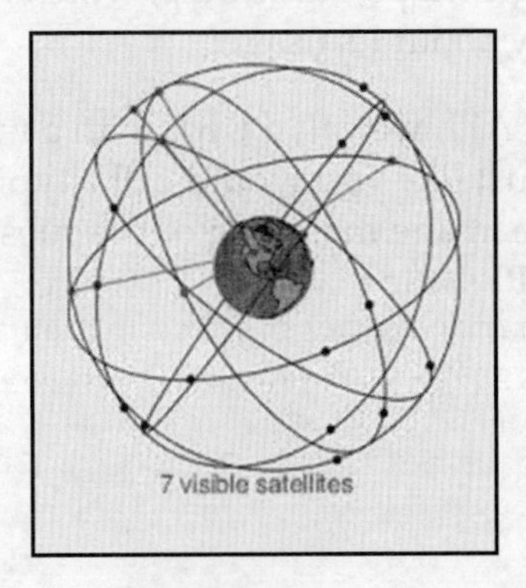

GPS Device tracking visible satellites

This is a list of terms we may come across when talking about or using GPS on a device:

- **Datum**: It is a term used in GPS to define the coordinate transformation system used to convert those satellite signals into usable coordinates. All GPS devices use WGS 84 as the standard, which is convenient for us since our maps also use WGS 84. A dedicated GPS device will support several different data according to an advanced user's needs.
- **Latitude/longitude**: By default, a GPS device will return latitude and longitude coordinates in WGS 84 datum. For us, this makes things easy, as we don't require any additional math conversions to map a device's location.
- **Altitude**: This represents the device's height above sea level. Most mobile GPS devices we will be building our game for do not support altitude, currently. Thus, we won't use altitude in the game, but hopefully this will we be supported in the future.
- **Accuracy**: This is reported by the device and represents the range of error calculated when determining the location. The more satellites a device acquires signals from, the better the location calculations will be. There is a limit to each device's accuracy and even what is possible on the public GPS network. A modern smartphone will often report at 5-8 meters in accuracy. However, some older smartphones can be as high as 75 meters. We will spend more time discussing GPS accuracy later in the book when we start to allow the player to interact with virtual objects.

Google Maps

As previously mentioned, we will use Google Maps for our map service. For this version of the game, we will use the Google Maps' static map, which means we do not need an API developer key or need to worry about its usage.

In order to use the static map's API, we call a **URL** with a number of query string parameters in a **GET** request, just like you would call a typical **REST** service. The **Google Maps API** then returns a single image matching our request. Here is an example of a request to the **Google Maps API** static map **REST** service:
`https://maps.googleapis.com/maps/api/staticmap?center=37.62761,-122.42588&zoom=17&format=png&sensor=false&size=640×480&maptype=roadmap.`

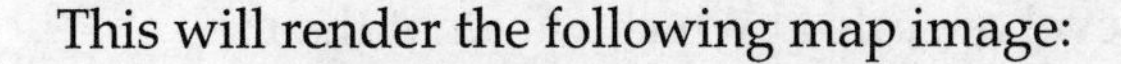

This will render the following map image:

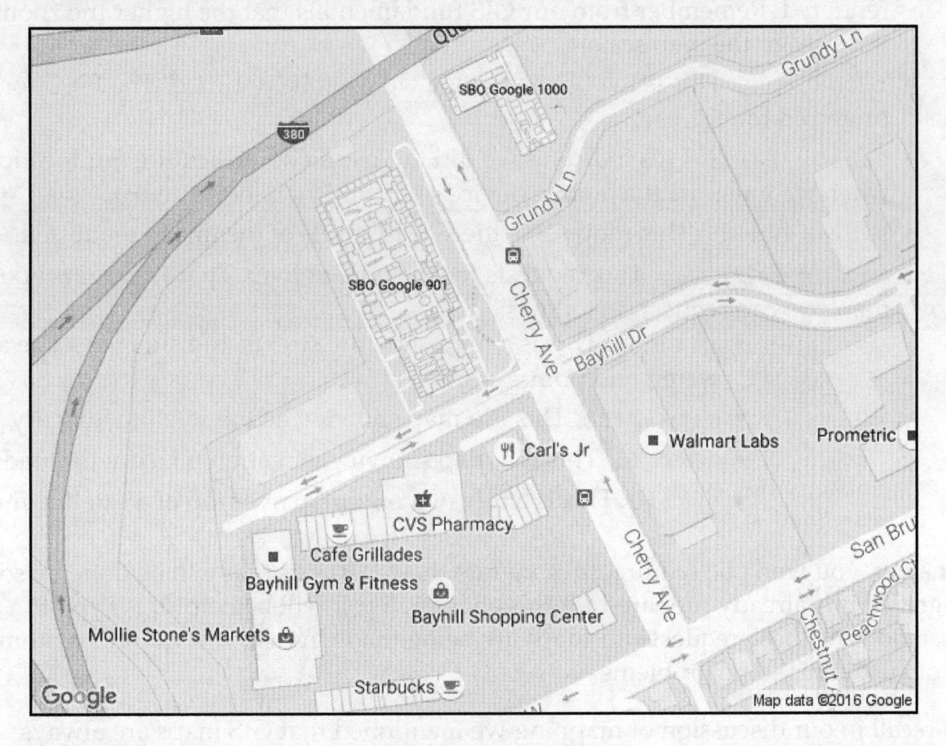

Image rendered from Google Maps API

Ensure that you test this on your own by clicking on the link or copying and pasting that URL into your favorite browser. Let's break down that URL into its component parts so we can understand the elements we need to define when requesting a map image:

- `https://maps.googleapis.com/maps/api/staticmap`: This represents the base URL of the Google Maps service. If we call this URL without any parameters, we will get an error. Let's take a look at each of the parameters and query syntax in some more detail.
- `?`: The question mark denotes the start of the query parameters.
- `center=37.62761,-122.42588`: This represents the center of our requested map in latitude and longitude coordinates.
- `&`: The ampersand symbol denotes the start of a new query parameter.

- `zoom=17`: This represents the zoom level or scale at which we want the map rendered. Remember from our GIS fundamentals that the higher the zoom level the smaller the map scale.
- `format=png`: This is the type of image we want be returned. The PNG format is preferred for our use.
- `sensor=false`: This indicates we did not use a GPS to acquire our location. Later when we integrate the mobile device's GPS, this will be set to true.
- `size=640x480`: This represents the size in pixels of the image requested.
- `maptype=roadmap`: This requests for the type of map. There are four types of map you can request, as follows:
    - `roadmap`: This is a map showing streets, transit, landscape areas, water, and points of interest.
    - `satellite`: This is a map showing actual satellite imagery.
    - `terrain`: This is a map showing elevation mixed with roadmap.
    - `hybrid`: This is a mix of a roadmap over top of a satellite map.

Fortunately, you won't be required to generate these URLs as that will be done by some packaged scripts already prepared for this chapter. Yet it will be helpful for you to understand how those requests for maps are being made in case you want to customize the game or encounter some problems.

If you recall in our discussion of mapping we mentioned that GIS maps are always constructed in layers. The great thing with Google Maps is that we can style our various map layers dynamically as part of the request. That allows us to style our maps specifically according to the game look and feel that we want. Take a few minutes to play with the Google Maps' style wizard available at `https://googlemaps.github.io/js-samples/styledmaps/wizard/`.

For our game, we have set a couple simple styles to give our game a darker look. The following screenshot shows the style selections as they are shown in the Google Maps style wizard:

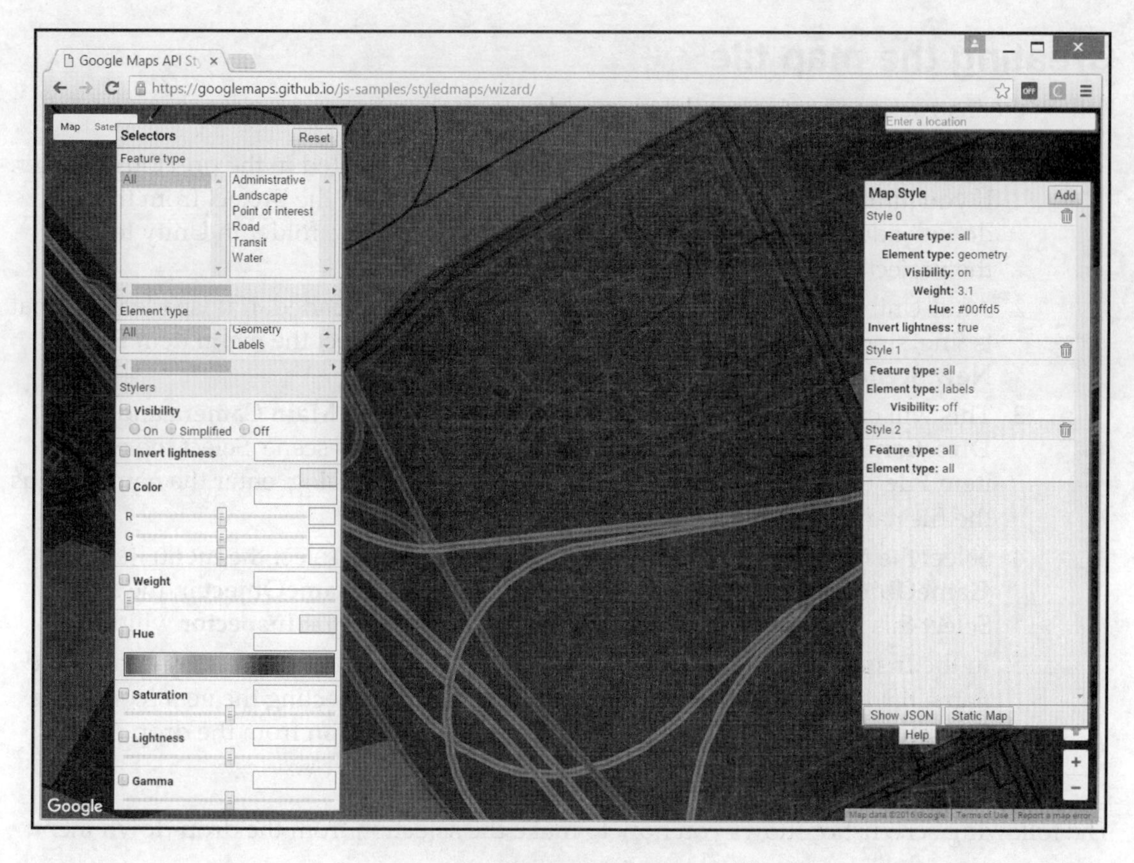

Game styles defined in the style wizard

For now, we won't get into extracting the styles from the wizard and using that to define our own map styles. That customization will be left for `Chapter 9`, *Finishing the Game*. Those readers who are curious can quickly see what those style parameters look like by clicking on the **Static Map** button on the left-hand panel in the style wizard.

Adding a map

After our brief introduction to mapping with GIS and GPS, let's jump back into Unity and add a map to our game. As we build out the map, we will review some of those GIS terms again. Let's continue where we left off in the previous chapter.

Creating the map tile

Follow the instructions here to get the map added to the game:

1. Open up Unity and load the FoodyGO project we created in the previous chapter. If you jumped ahead to this chapter, you can also load the project from the downloaded source code. Open the `Chapter_2_Start` folder in Unity to load the project.
2. After Unity opens, you should see the **Splash** screen loaded. If it isn't loaded, that is fine, since we will be creating a new map scene. Select the menu item **File** | **New Scene**.
3. This will create new empty scene in Unity with only a **Main Camera** and **Directional Light**. Before we forget, let's save this new scene. Select the menu item **File** | **Save Scene as...**, and on the **Save Scene** dialog, enter the name `Map` as the file name and then click on **Save**.
4. Select the **Map** scene in the **Hierarchy** window. Then, select the menu item **GameObject** | **Create Empty** to create a new empty **GameObject** in the scene. Select this new game object and view its properties in the **Inspector** window.
5. In the **Inspector** window, rename the **GameObject** to `Map_Tiles` by editing the name field. Reset the object's transform position by selecting the gear icon in the **Transform** component and then selecting **Reset Position** from the drop-down menu.

The following screenshot shows you how to make the selection from the drop-down menu:

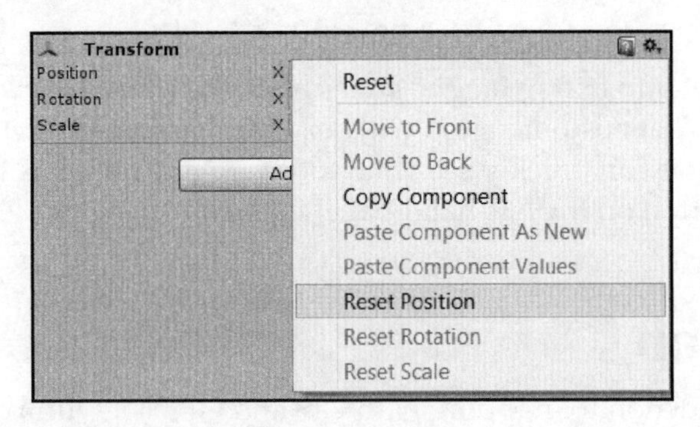

Resetting a game object's position

1. We will generally reset most of our game objects to or near zero in order to simplify any GIS math conversions. The **Map_Tiles** game object should now have a zero or identity transform as shown in the following screenshot:

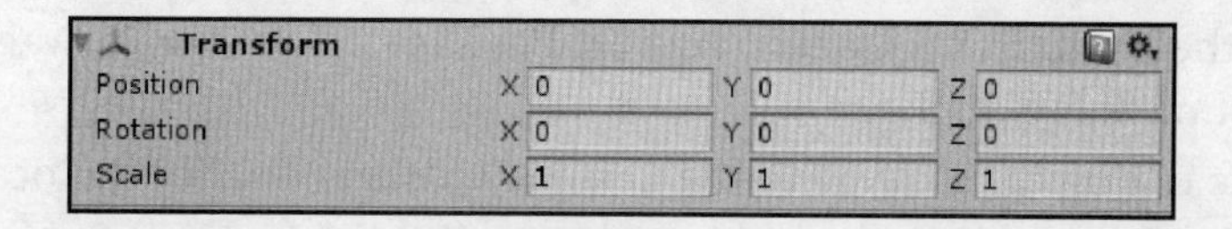

A game object with zero transform

2. With the **Map_Tiles** game object selected, right-click (press *command* and click, on a Mac) to open the context menu and select **3D Object | Plane**. Here is a screenshot showing you how to add the plane from the context menu:

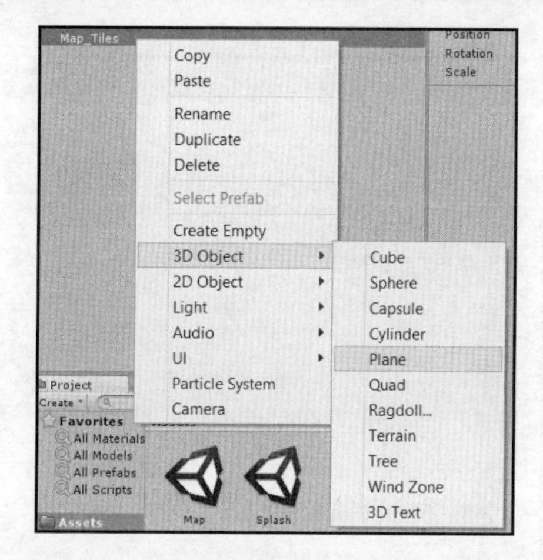

Game object's context menu

3. Select the plane game object and rename it to `Map_Tile` in the **Inspector** window. Make sure that the object's transform is zero.
4. Double-click on the **Map_Tile** plane in the **Hierarchy** window to focus on the object in the **Scene** window. If the object is not visible, make sure that the 2D button is turned off in the **Scene** window.
5. In the **Inspector** window, edit the **Transform** component properties and then edit the scale of the **X** and **Z** to `10`. Notice how the plane's dimensions expend as you edit the scale.

6. We now need to add a script to our **Map_Tile** object that will render our map. At this point, we will avoid creating new scripts and instead just add script from an imported asset. Of course, later in the book, we will be building new scripts. The scripts we need will be in the downloaded code in `Chapter_2_Assets` folder. Open the menu item **Assets** | **Import Package** | **Custom Package...** to open the **Import package** dialog.
7. Use the dialog to navigate to the `Chapter_2_Assets` folder located in the downloaded source code folder and import the `Chapter2.unitypackage` asset by clicking on **Open**.
8. A progress dialog will appear that will show loading of the assets, which will quickly be replaced by the **Import Unity Package** dialog. Make sure that all the items in the dialog are selected and available to import and then click on **Import**. The **Import Unity Package** dialog with the scripts being imported will appear:

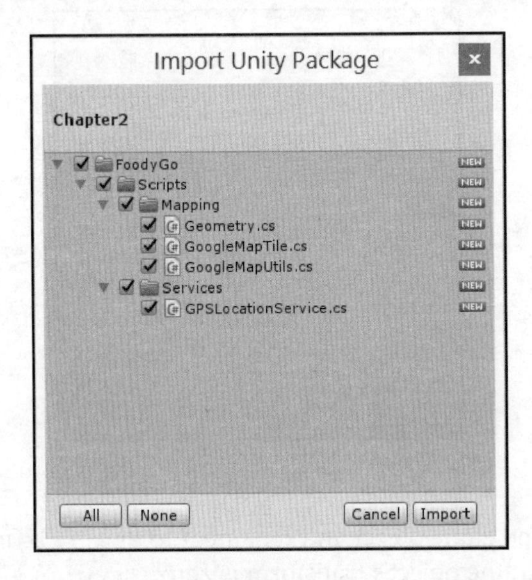

Chapter assets being imported

9. After you import the assets, you will notice a new folder created under **Assets** in the **Project** window. Feel free to explore this new folder and its contents so you are familiar with how the project assets will start to be organized. Notice how the folder contents match exactly what we imported.

10. Select the **Map_Tile** object in the **Hierarchy** window. Click on the **Add Component** button seen in the bottom of the **Inspector** window. A context menu will open showing you a list of components; select **Mapping** | **Google Tile Map**. This will add the Google mapping script component to the **Map_Tile** game object.

    Google Maps for Unity available for free on the Unity Asset Store was the inspiration for the **Google Tile Map** code. Several things were changed for the more advanced version used in the game.

11. In the **Inspector** window, edit the **Google Map Tile** script to match the values, as follows:

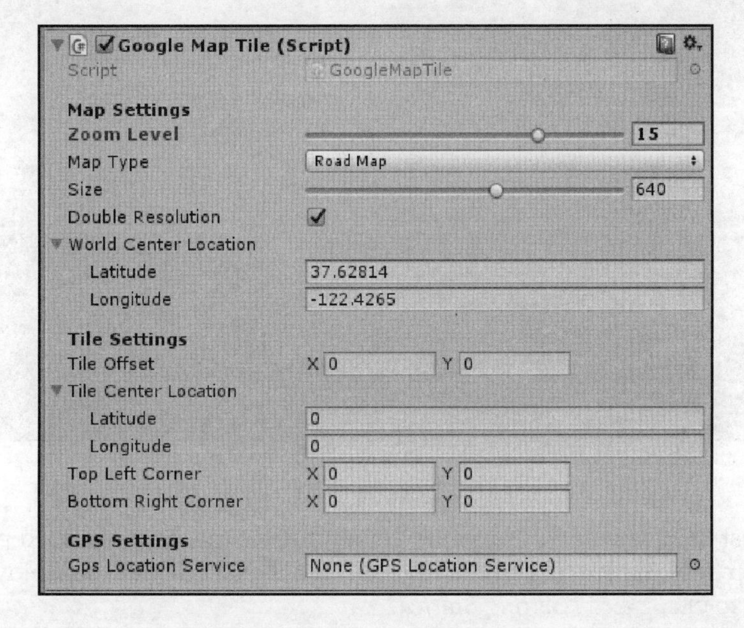

Editing the component values to match the screen

12. Hopefully, those mapping parameters make some sense after the preceding introduction. For now, we will use a **Zoom Level** of **15** to test how our make works. Those location coordinates are for Google located in San Francisco. Of course, we will connect the device's GPS and use your local coordinates later.

13. Press the Play button. After a couple of seconds, your screen should look similar to this:

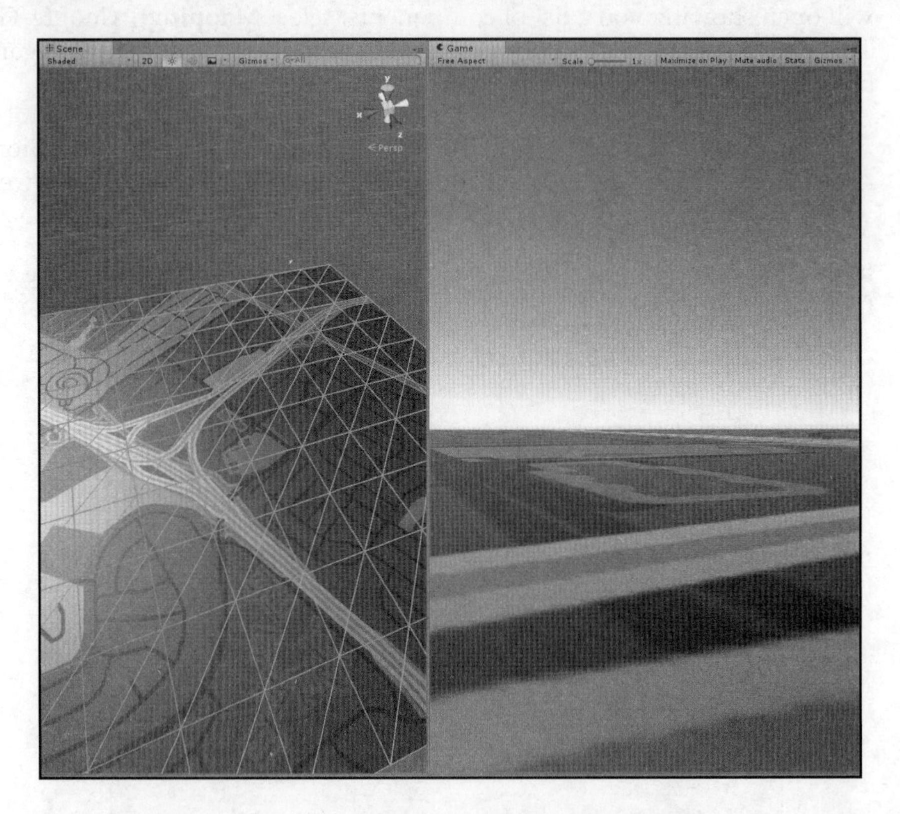

Google Map running within Unity

14. To test the game on your mobile device, follow the steps we used previously to set up deployment. If you are unclear on the exact details of deploying the game, refer to `Chapter 1`, *Getting Started*.

Here is a quick summary of the steps, assuming you have previously deployed the game to your device:

1. Make sure that your device is connected to the Unity development machine by a USB cable.
2. Save your scene by choosing **File | Save Scene**.
3. Save your project by selecting **File | Save Project**; note it is always a good habit to save your game before building. Unity build is notorious for crashing the editor.

4. Select the menu item **File** | **Build Settings...** to open the **Build Settings** dialog.
5. Uncheck the **Splash** scene, as we don't need it yet. Click on the button **Add Open Scenes** to add the **Map** scene to the build.
6. Select your appropriate deployment platform, Android or iOS.
7. Click on the **Build and Run** button to start the build and deployment process.
8. When prompted, save and overwrite your deployment to the same location as you chose before.
9. Wait for the game to finish building and then deploy it onto your device.
10. After the game loads on your device, inspect the map and rotate your device. At this point, the game does very little, but this indeed shows that the map is functioning on your device.

The first thing you may notice is that the map image is brighter. This brightness is caused by our lighting and the default material on the plane. Fortunately, for us, this visual style is what we are after and we will leave the added brightness as it is.

Secondly, you will notice that the map is more pixelated than the image we rendered on the server in the preceding style wizard. That pixelation is a result of stretching our map image across the plane. The obvious solution is to increase the image size and resolution. Unfortunately, the maximum image size we can request from Google Maps is around 1200 x 1200 pixels, which is what we are already doing using double resolution. This means we need to find a different solution to get a cleaner, crisp-looking map. In the following section, we will resolve this pixelation issue.

Laying the tiles

Due to the level of detail on maps and the fine lines, we generally always want to render our maps at the highest resolution possible. Unfortunately, rendering high-resolution images is performance intensive and prone to errors. Fortunately, there are plenty of examples on how others have solved this resolution issue in mapping by stitching multiple images or image tiles together.

We will take the exact same approach and extend our map from a single tile to a 3 x 3 grid of tiles, as follows:

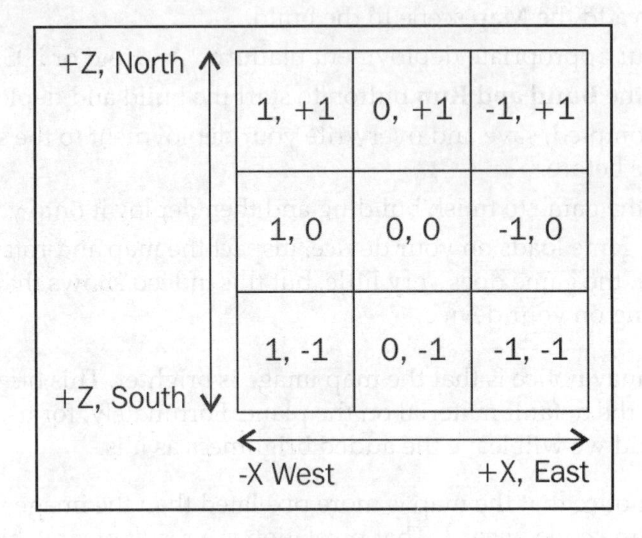

Map tile layout for a 3 x 3 grid

Note that in the diagram, the *x* axis and the tile offsets are inversely related, or in other words, a tile offset of 1 in the X direction will need to be offset on the *x* axis in 3D space in the negative direction. The *z* axis and Y tile offset are in a direct relationship. This means that as the Y tile offset is set to 1, the *z* axis will also be set a positive value. As our player is close to the ground, the game only needs a 3 x 3 grid. If you build a game with a higher-level camera or want to show farther in the horizon, you would extend the tile layout to 5 x 5, 7 x 7, 9 x 9, or whatever size you need.

So, let's get started and extend our map from a single tile to the 3 x 3 grid tiles. Follow these instructions in Unity to build the map tile layout:

1. Select the **Map_Tile** game object in the **Hierarchy** window. In the **Inspector** window, edit the object's properties to match these values:
    - **Transform**, **Scale**, **X**: 3
    - **Transform**, **Scale**, **Z**: 3
    - **Google Map Tile**, **Zoom Level**: 17

2. Create a new folder under the `FoodyGO` folder by right-clicking (Press *command* and click on Mac) to open the context menu and select item **Create** | **New Folder**. Rename the folder in the highlighted edit window to `Prefabs`.

3. We will now create a prefab of our **Map_Tile** object. You can think of prefabs as game object copies or templates. To create the prefab, select and drag the **Map_Tile** game object to the new `Prefabs` folder we just created. You will see a new prefab appear in the folder called **Map_Tile**. After the prefab is created, you will notice that the **Map_Tile** game object turns blue in the **Hierarchy** window. The blue highlight means the game object is bound to a prefab.
4. Go back to the **Hierarchy** window and again select the **Map_Tile** object and rename it `Map_Tile_0_0`. We did this in order to denote this as the center or 0,0 tile.
5. Select the **Map_Tile_0_0** game object in the **Hierarchy** window and enter *Ctrl* + *D* (*command* + *D* on a Mac) to duplicate the map tile. Do this eight times to create eight additional map tiles, as follows:

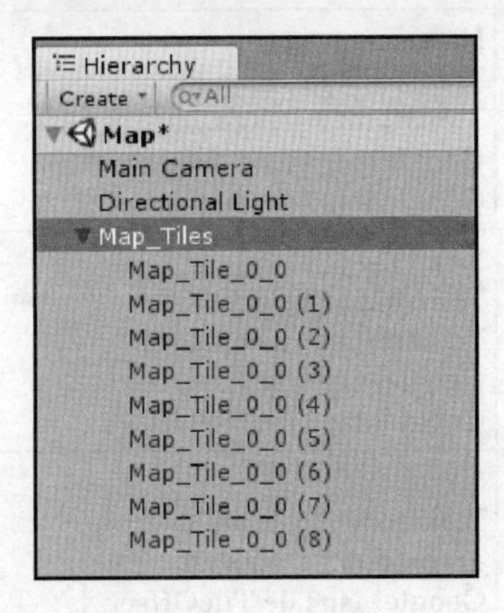

Copied map tiles shown parent to the Map_Tiles game object

6. Rename the copied map tiles and set the properties of each within the **Inspector** window, as shown in the following table:

Game object	Properties
Map_Tile_0_0 (1)	**Name:** `Map_Tile_0_1` **Transform.Position.X:** `0` **Transform.Position.Z:** `30` **GoogleMapTile.TileOffset.X:** `0` **GoogleMapTile.TileOffset.Y:** `1`
Map_Tile_0_0 (2)	**Name:** `Map_Tile_0_-1` **Transform.Position.X:** `0` **Transform.Position.Z:** `-30` **GoogleMapTile.TileOffset.X:** `0` **GoogleMapTile.TileOffset.Y:** `-1`

Map_Tile_0_0 (3)	**Name**: `Map_Tile_1_0` **Transform.Position.X**: `-30` **Transform.Position.Z**: `0` **GoogleMapTile.TileOffset.X**: `1` **GoogleMapTile.TileOffset.Y**: `0`
Map_Tile_0_0 (4)	**Name**: `Map_Tile_-1_0` **Transform.Position.X**: `30` **Transform.Position.Z**: `0` **GoogleMapTile.TileOffset.X**: `-1` **GoogleMapTile.TileOffset.Y**: `0`
Map_Tile_0_0 (5)	**Name**: `Map_Tile_1_1` **Transform.Position.X**: `-30` **Transform.Position.Z**: `30` **GoogleMapTile.TileOffset.X**: `1` **GoogleMapTile.TileOffset.Y**: `1`
Map_Tile_0_0 (6)	**Name**: `Map_Tile_-1_-1` **Transform.Position.X**: `30` **Transform.Position.Z**: `-30` **GoogleMapTile.TileOffset.X**: `-1` **GoogleMapTile.TileOffset.Y**: `-1`
Map_Tile_0_0 (7)	**Name**: `Map_Tile_-1_1` **Transform.Position.X**: `30` **Transform.Position.Z**: `30` **GoogleMapTile.TileOffset.X**: `-1` **GoogleMapTile.TileOffset.Y**: `1`
Map_Tile_0_0 (8)	Name: `Map_Tile_1_-1` **Transform.Position.X**: `-30` **Transform.Position.Z**: `-30` **GoogleMapTile.TileOffset.X**: `1` **GoogleMapTile.TileOffset.Y**: `-1`

7. Press the play button to run the game. While the game is running, select the **Map_Tile_0_0** in the **Hierarchy** window and type `F` , to frame the object in the **Scene** window. Notice how the pixelation on the map has been dramatically reduced. You should see something similar to the following screenshot:

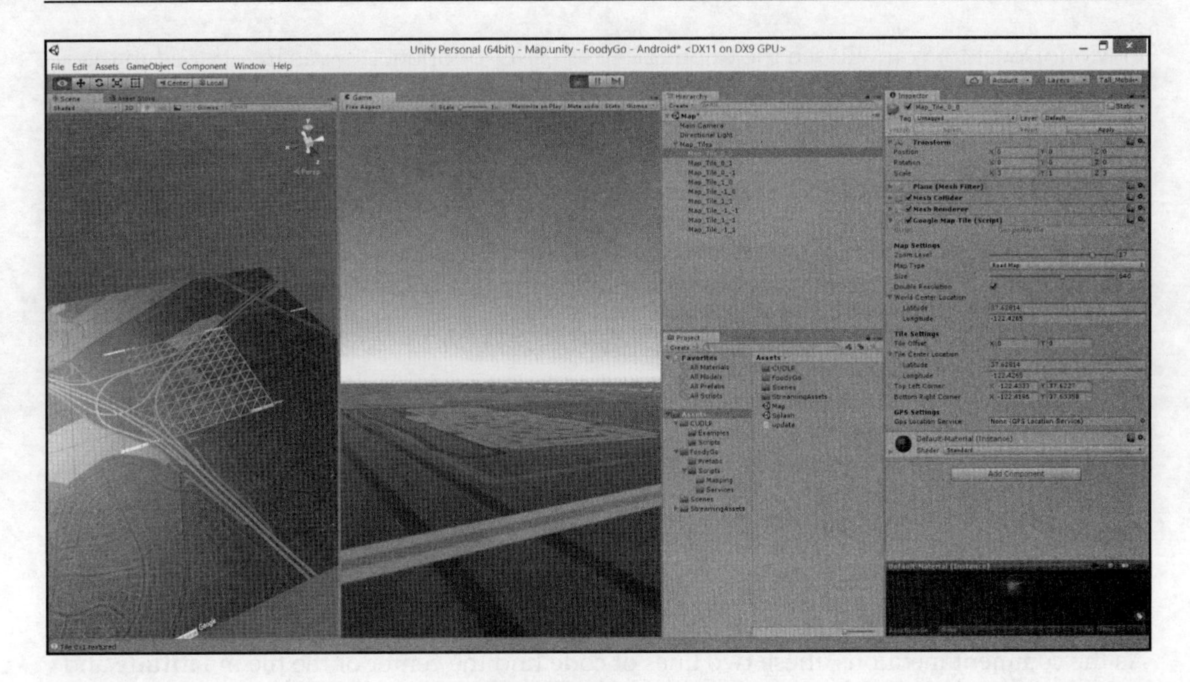

Game running in the play mode showing our tiled map

Understanding the code

Great, we now have a cool-looking map in our game. Of course, the process was a bit repetitive to build the map; but if you are careful, it should not take too long. If you were expecting a lot of math in order to line up those tiles, fortunately that is all done in the `GoogleMapTile` script. Let's take this opportunity to take a break from Unity and look at the `GoogleMapTile` script in **MonoDevelop**.

In Unity, select **Map_Tile_0_0** in the **Hierarchy** window. Go to the Inspector window, and on the **Google Map Tile** script component, click on the gear icon to open the context menu. From the menu, select **Edit Script**. You will see a progress bar, and after a few seconds, MonoDevelop will open.

MonoDevelop is the default script editor for Unity. If you are developing on Windows, Visual Studio Community or above is also an excellent option. Another good option is Visual Studio Code, which is a lightweight alternative for Windows, Mac, and Linux.

In MonoDevelop, you will see the `GoogleMapTile` script open. As was mentioned in the book's prerequisites, you should have a basic knowledge of C#, so the contents of the script should not look overly intimidating. If you are new to Unity scripting, that is fine, as we will get into more detail about writing scripts later. For now, we will concentrate on a few areas of the code that will show how the map tiling works.

Scroll down through the code until you reach the method, `IEnumerator _RefreshMapTile()`. Here is an excerpt from the top lines of that method we will look at in more detail:

```
IEnumerator _RefreshMapTile ()
      {
            //find the center lat/long of the tile
            tileCenterLocation.Latitude =
GoogleMapUtils.adjustLatByPixels(worldCenterLocation.Latitude,  (int)(size
* 1 * TileOffset.y), zoomLevel);
            tileCenterLocation.Longitude =
GoogleMapUtils.adjustLonByPixels(worldCenterLocation.Longitude, (int)(size
* 1 * TileOffset.x), zoomLevel);
```

As the comment mentions, these two lines of code find the center of the tile in **latitude** and **longitude** map coordinates. They do that by taking the tile image size (`size`) and multiplying that by the `TileOffset.y` for latitude and `TileOffset.x` for longitude. The result of that multiplication and the `zoomLevel` is passed to a `GoogleMapUtils` helper functions to calculate the adjusted **latitude** or **longitude** for the tile. Seems simple, right? Of course, the bulk of the work is done in the `GoogleMapUtils` functions, which are just standard GIS math functions for converting distances. If you are curious, take a look at the `GoogleMapUtils` code, but for now we will continue looking at just the `_RefreshMapTile` method.

Continue scrolling down through the code until you come to this excerpt:

```
//build the query string parameters for the map tile request
queryString += "center=" + WWW.UnEscapeURL (string.Format ("{0},{1}",
tileCenterLocation.Latitude,  tileCenterLocation.Longitude));
queryString += "&zoom=" + zoomLevel.ToString ();
queryString += "&size=" + WWW.UnEscapeURL (string.Format  ("{0}x{0}",
size)); queryString += "&scale=" + (doubleResolution ? "2" : "1");
queryString += "&maptype=" + mapType.ToString ().ToLower ();
queryString += "&format=" + "png";

//adding the map styles
queryString +=
"&style=element:geometry|invert_lightness:true|weight:3.1|hue:0x00ffd5";
queryString += "&style=element:labels|visibility:off";
```

As the comment describes, this section of the code is what builds up those query parameters that are passed to the Google Maps API to request a map image. Since we are passing these parameters in a URL, we need to make sure that you encode special characters and that is what the `WWW.UnEscapeURL` calls do. Notice that at the bottom we are also adding a couple styles. In `Chapter 9`, *Finishing the Game*, we will take a look at how you can easily add your own styles using the **Google Maps Style Wizard**.

Finally, scroll down to the bottom of the `_RefreshMapTile` method; the following is an excerpt of the code:

```
//finally, we request the image
var req = new WWW(GOOGLE_MAPS_URL + "?" + queryString);
//yield until the service responds
yield return req;
//first destroy the old texture first
Destroy(GetComponent<Renderer>().material.mainTexture);
//check for errors
if (req.error != null)
{
      print(string.Format("Error loading tile {0}x{1}:  exception={2}",
                  TileOffset.x, TileOffset.y, req.error));
}
 else
 {
      //no errors render the image
      //when the image returns set it as the tile texture
      GetComponent<Renderer>().material.mainTexture = req.texture;
         print(string.Format("Tile {0}x{1} textured", TileOffset.x,
TileOffset.y));
        }
```

In the first line, the code uses the `WWW` class to make a request to the `GOOGLE_MAPS_URL` appended with the earlier constructed `queryString`. The `WWW` class is a Unity helper class that allows us to make calls to URLs for virtually anything. Later in the book, we will use this class to make other service requests.

The next line, `yield return req;`, essentially tells Unity to continue on until this request responds. We can do that here because this method is a coroutine. **Coroutines** are methods that return **IEnumerator** and are an elegant way to prevent thread blocking. If you have ever done more traditional C# asynchronous programming, you will certainly appreciate the beauty of coroutines. As before, we will cover more details about coroutines when we get into script writing.

Next, we call `Destroy` on the object's current texture. `Destroy` is a public method of the `MonoBehaviour` class that safely allows us to destroy objects and all components attached to the object. If you are seasoned C# Windows or Web developer, this step may seem quite foreign to you. Just remember that we have to be mindful of memory management that can quickly get out of hand when running a game. In this example, if we were to remove this line of code, the game would likely crash due to texture memory leaks.

After the `Destroy` call, we do an error check just to make sure that no errors occurred while requesting the image tile. If an error occurs, we just print an error message. Otherwise, we swap the current texture for a new downloaded image. We then use `print`, in order to write a debug message to the **Console** window. The `print` method is the same as calling `Debug.log`, but is only available from a class derived from `MonoBehaviour`.

Our final look at the code will be to understand when the `_RefreshMapTile` method is called. Scroll up through the code until you find the `Update` method, as follows:

```
// Update is called once per frame
        void Update ()
        {
            //check if a new location has been acquired
            if (gpsLocationService != null &&
                gpsLocationService.IsServiceStarted &&
                lastGPSUpdate < gpsLocationService.Timestamp)
            {
                lastGPSUpdate = gpsLocationService.Timestamp;
                worldCenterLocation.Latitude =
gpsLocationService.Latitude;
                worldCenterLocation.Longitude =
gpsLocationService.Longitude;
                print("GoogleMapTile refreshing map texture");
                RefreshMapTile();
            }
        }
```

`Update` is a special Unity method available on every `MonoBehaviour` derived class. As the comment mentions, the `Update` method is called every frame. Obviously, we don't want to refresh the map tile every frame since it is unlikely the request would return that quickly anyway. Instead, we would first want to make sure that we are using a location service and it has started. Then, we check whether the location service has detected movement by checking a timestamp variable. If it passes those three tests, then we update the timestamp, get a new world center, print a message, and finally call `RefreshMapTile`. `RefreshMapTile` makes a call to `StartCoroutine(_RefreshMapTile)` that starts the tile refresh.

Since we haven't started connecting the GPS service yet, this all may seem foreign. Not to worry, we will get to that shortly, but for now it will be helpful to understand how frequently the map tiles will be redrawn.

In this section, we enhanced the resolution of our game map by rendering image tiles rather than a single image. For our purpose, we still used a fairly large tile size for each map tile image. We can get away with this because our camera will be above the player looking down. As you can see though, it is a simple process to create any size of tile map. If you do decide to create a larger map, just be aware that downloading several map tiles could increase a player's data consumption dramatically.

Setting up services

Services can have a broad definition depending on your application and need. For us, we will use the term services to denote any code that runs as a self-managed class that other game objects consume. Services are different than a library or global static class, such as the `GoogleMapUtils` class, because they run as an object or objects. In some cases, you may decide to implement services using the singleton pattern. As for this book, our intention is to write simpler code, so we will create and use services as game objects.

For this chapter, we will set up two services. The GPS Location Service, for finding the player's location and CUDLR, for debugging. Let's start by getting the CUDLR service started, as that will help us in debugging any issues we may have when we set up the location service.

Setting up CUDLR

CUDLR stands for **Console for Unity Debugging and Logging Remotely** and is a free asset available from the Unity Asset Store. We will use CUDLR to not only watch the activity of our device as the game plays but also execute some simple console commands remotely. We will look at Unity Remote, which is another diagnostic tool, in the following chapter. It is very powerful, but can be problematic to run and often fails to access the location service, even though Unity claims this is supported. As you get more into the development of the game, you will see that it is always helpful to have a remote method of monitoring and controlling our game.

In order to use CUDLR, your device and development machine must be on the same local Wi-Fi network. Skip this section if your mobile device is not able to connect to your local Wi-Fi network.

Perform the following steps to install and set up CUDLR:

1. Open the Asset Store window by selecting menu item **Window | Asset Store**. After the window opens, enter `cudlr` in the **Search** field and press Enter. After a few seconds, the asset list should appear.
2. Click on the image or link for CUDLR to load the asset page. After the page loads, you will see an **Import** button; click on that button to import the asset into the project.
3. The asset is quite small, so it should download quickly. After it finishes downloading, you will see the **Import Unity Package** dialog. Just make sure that everything is selected, as shown in the following screenshot:

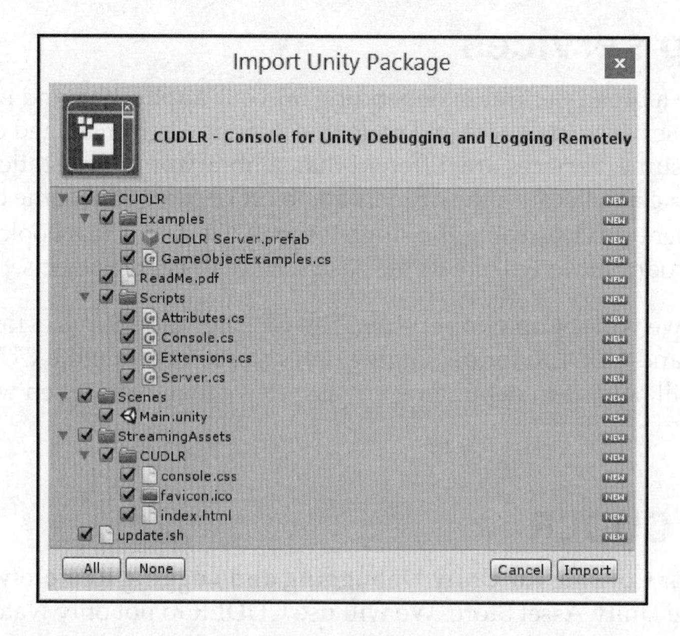

Importing the CUDLR asset

4. Since we are importing this asset into a project for the first time, we are going to install everything. Later on in development, we can always decide to remove unneeded parts of an asset, such as the **Examples** folder. When you are ready to import, click on the **Import** button on the dialog and wait for the import to finish.
5. Create a parent services object in the scene by selecting menu item **GameObject | Create Empty**. This will create a new empty game object in the **Hierarchy** window. Rename the new object **Services**.

6. Right-click (press *Ctrl* and click on a Mac) on the **Services** object in the **Hierarchy** window to open the context window. From the context menu, select **Create Empty**. This will create an empty object called **GameObject** as a child of the **Services** object. Repeat the process to create another empty game object.
7. Select the first empty **GameObject** and rename it `CUDLR`. Then, select the second and rename it `GPS`. We will add the GPS service later, but since we are here let's be efficient.
8. Open the **Assets/CUDLR/Scripts** folder in the **Project** window. Select and drag the server script from the **Project** window onto the **CUDLR** game object in the **Hierarchy** window. This will add the CUDLR server component to the game object. That's it, CUDLR is ready to go.

Debugging with CUDLR

What makes CUDLR such a useful tool is that it turns part of our game into web server; yes, a web server. We can then view and communicate with our game just as if we had installed a backdoor. Since CUDLR is accessible to any computer on the network, we also don't need to have a physical connection or even run Unity to control the game. Of course, having your game run as a local web server is a security risk to your game and possibly your player's device. So, before you ship your game, just delete the CUDLR service game object to deactivate it.

Follow the instructions here to connect to the CUDLR service running in your game:

1. Open you mobile device and find the IP address, write this down or remember it. Perform the following steps to find the IP address for Android or iOS:
    - **Android**: Go to **Settings** | **About phone** | **Status** and scroll down to the **IP address** field.
    - **iOS**: Select **WiFi** to open a list of currently available wireless networks. Locate the wireless network to which you are currently connected in the list, then tap on the blue circle with a white arrow that appears to the right of the network name. The **IP Address** field should be the first field listed under the **DHCP** tab.
2. Build and deploy the game to your mobile device using the steps previously mentioned in this chapter. Make sure that the game is open and running on your device.
3. Disconnect your USB cable from your device and computer.
4. Open a web browser; Chrome is a good choice. Type the following URL in your address bar: `http://[Device IP Address]:55055/`.

5. You should see the CUDLR console window in your browser, with output very similar to this screenshot:

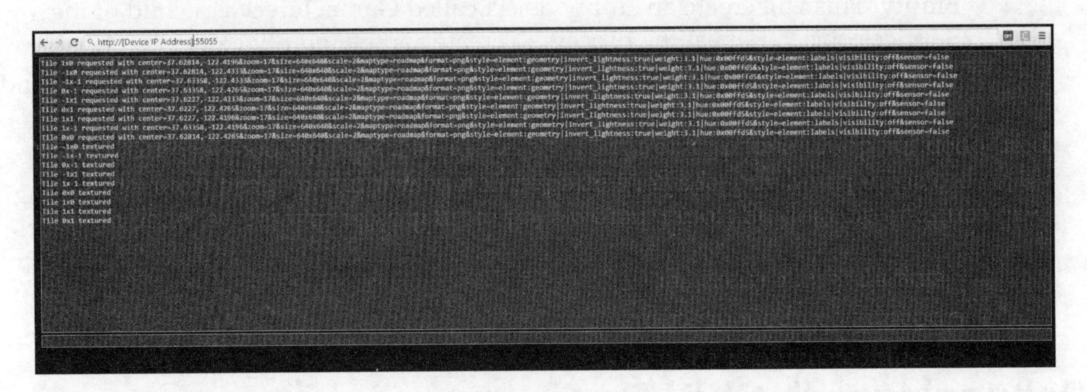

Browser showing CUDLR service console

6. There should be nine request calls and nine response calls being logged in the window as the tile map is loaded. Hopefully, by now, you can read and understand a little of what those requests do.
7. So, what else can the console do? Type `help` in the bottom textbox below the console output. That will list all the commands available. Currently, there are only a few commands, but we will add to those later. If for some reason, the console doesn't respond, try to refresh the page and/or make sure that the game is running on your mobile device.

If you have problems running or connecting to CUDLR, please refer to `Chapter 10`, *Troubleshooting*. As we mentioned previously, we will look at other options for debugging and diagnosing issues while developing with Unity. However, CUDLR, which can run completely remotely, will be our best option for testing our game, as we test real-world movement and GPS tracking. Speaking of GPS, time to finish up with the final section of this chapter and bring everything together.

Setting up the GPS service

The final piece we need to generate a real-world location-based map is to find where the device is located. Of course, the best way, as we learned in the previously section, is to use the built-in GPS to determine where the device is located in latitude and longitude coordinates. Like the tile map, we will use an imported script to build the service and get us running quickly without getting into any scripting.

Before we begin, be sure that your device has the location service enabled by checking the following:

- **Android**: Go to **Settings** | **Location** and confirm that the service is on.
- **iOS**: Follow these instructions:

    1. Tap on **Privacy** | **Location Services**.
    2. Scroll down and tap on **FoodyGO**.
    3. Decide whether to allow location access **Never** or **While Using the App**.

Now, perform the following instructions to install the GPS service code and test the game:

1. Select the **GPS** game object under **Services** in the **Hierarchy** window.
2. Click on the **Add Component** button in the **Inspector** window, and from the component list, select **Services** | **GPSLocationService**.
3. You will see the **GPSLocationService** component get added to the **GPS** service object.
4. Select **Map_Tile_0_0** in the **Hierarchy** window and then press *Shift* and click (same for Mac) on the bottom map tile in the list to select all the nine map tiles together.
5. With all the nine map tiles selected, drag the **GPS** service object in the **Hierarchy** window to the **Gps Location Service** field, as follows:

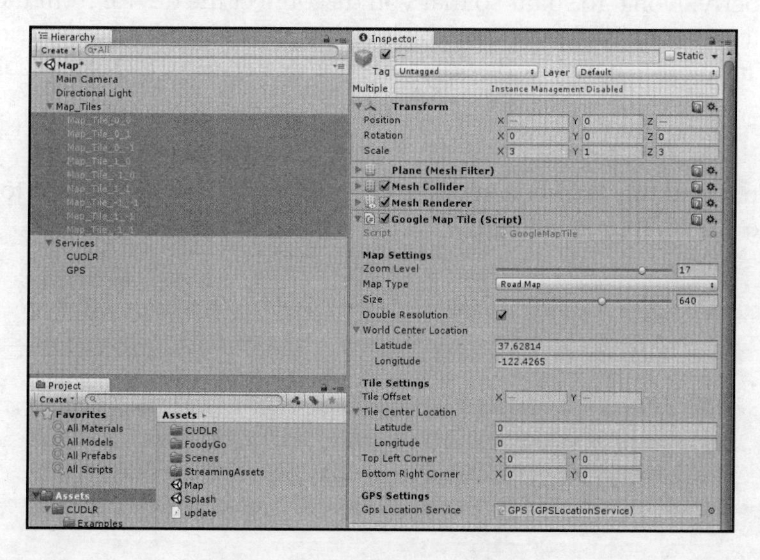

Editing multiple map tile objects at the same time

6. What we just did is edit all the nine map tiles at the same time and add the **GPSLocationService** to each. Remember that in our investigation of the `GoogleMapTile` script, we had noticed that the map tiles will call the `GPSLocationService` in order to find the map world center coordinates.
7. Now that the GPS service is connected to all the map tiles, press Play and let's see what it looks like.
8. If you are scratching your head wondering what went wrong, not to fear. Nothing has actually gone wrong. The problem is that the Unity editor has no access to a location service or GPS while running on your computer. What we need to do is deploy the game to your device.
9. Build, deploy, and run the game on your device just as you have previously done. You should be feeling comfortable deploying the game now.
10. You should now see a close-up map of the area around you. The map will likely look offset, and that is due to the position of the camera. Don't worry, we will fix that in the next chapter. If you are having trouble seeing a different map, then just verify each of the above steps. For those who continue to have issues even after confirming the previous steps, refer to `Chapter 10`, *Troubleshooting*.
11. Ensure that you go back to your browser and refresh you CUDLR console page. Note all the console outputs we are seeing now. Pay special attention to the map tile requests being made; those center coordinates should now match your location coordinates.
12. Unplug your device from your computer and wander around your house or property. Don't go so far so that you disconnect the device from the network, but try to get your GPS to update the location. Perhaps even have a friend wander around with the mobile device while you watch the CUDLR console.

Have fun...

Hopefully, you found this last section of the chapter to be a rewarding end to a very quick introduction to GIS, mapping, and GPS.

Summary

In this chapter, you were introduced to some fundamentals about GIS, mapping, and GPS. This basic knowledge helped us define some terminologies for working with and loading the Google Maps API in Unity. We then added a map to the game but decided the quality was lacking. This led us to building a tile map system for the game map. After which, we took a quick break to introduce a console debugging tool called CUDLR. CUDLR helped us debug a fundamental piece of our game and find the player's location via GPS. This allowed us to finish out the chapter by adding GPS to our game using the GPS location service.

Now that we have the fundamentals established, we can now get into more hands-on game development. The following chapter will be a whirlwind introduction to adding a fully rigged character to our scene and also cover mobile touch input, free look cameras, and how to access the device's motion sensors.

3

Making the Avatar

Every game needs some element that denotes the player's presence and point of interaction with the game's virtual world. This element could be a car in a racing game, a worm in a maze game, hands and a weapon in an FPS, or an animated character in an adventure or role-playing game. We will use the latter, an animated character, to denote our player's avatar and location in the game using a third person-perspective camera to showcase the character's movement, around the real world map. This should give our players a good immersive game experience that will make our game fun to play.

Unlike the previous chapters where we spent some time discussing the background and the terminology, in this chapter, we will dive right into Unity and start to add the player avatar. In this entire chapter, we will discuss new game development concepts and cover the following topics:

- Importing standard Unity assets
- 3D animated characters
- Third-person controller and camera positioning
- Free look camera
- Cross-platform input
- Creating a character controller
- Updating component scripts
- Using an iClone character

We will continue developing from where we left off in the last chapter. So, open up the `FoodyGO` project in Unity and let's get started. If you have jumped ahead from a previous chapter, open the downloaded `Chapter_3_Start` project with the book's source code.

Importing standard Unity assets

Game development is a complex undertaking that requires a keen understanding of the hardware platform, graphics rendering, and managing game assets. Unity makes all that substantially easier by building a cross-platform game engine that abstracts many of those complex details away from the developer. However, since no two games are exactly alike, Unity also supports extensibility through importing assets and plugins. Assets may include everything from scripts and shaders to 3D models, textures, and sound. The ability to quickly extend a game through assets is a powerful feature in Unity and is something we will cover extensively in this chapter.

Let's get started by importing some of the standard Unity assets into our game project. Unity provides a number of standard or reference assets that developers can freely use in their game. Using these standard assets is often a great way to quickly start development. However, due to particular elements of game, design or visual esthetics need to be rewritten or replaced. For now, we will follow the same pattern and start with the standard assets, but we will likely need to rewrite or replace some elements later.

Perform the following instructions to import the standard assets we will use in this chapter:

1. Open the project in Unity and make sure that the **Map** scene is loaded. Again, if you have moved ahead from another chapter, open the `Chapter_2_End` folder from the downloaded source code as a project in Unity.
2. Select the menu item **Assets** | **Import Package** | **Cameras**. After a short time, the package will be downloaded, and you will see the **Import Unity Package** dialog, as shown in the following screenshot:

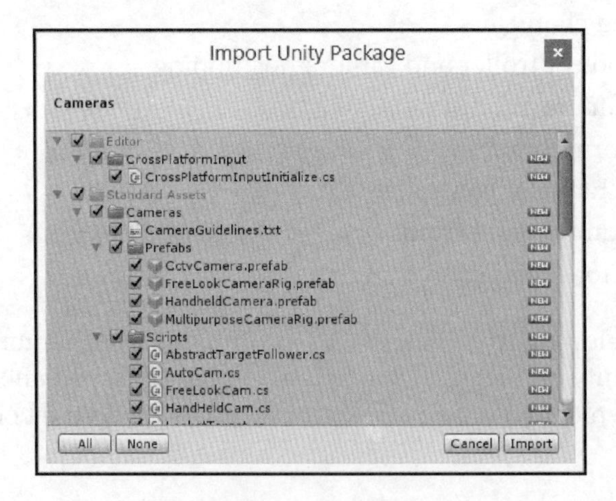

Importing the camera's Unity standard asset

3. Ensure that all the items are selected and then click on the **Import** button. After the asset package finishes importing, you will see a new folder called **Standard Assets** in your `Assets` folder in the **Project** window. If you open the new folder, you may also notice that **CrossPlatformInput** was also added. That is typical of Unity and other assets and is something to be aware of. For now, let's not worry about it.
4. Now, let's import the **Characters** asset by selecting the menu item **Assets** | **Import Package** | **Characters**. After a short time, the package will be downloaded, and you will see the **Import Unity Package** dialog, as follows:

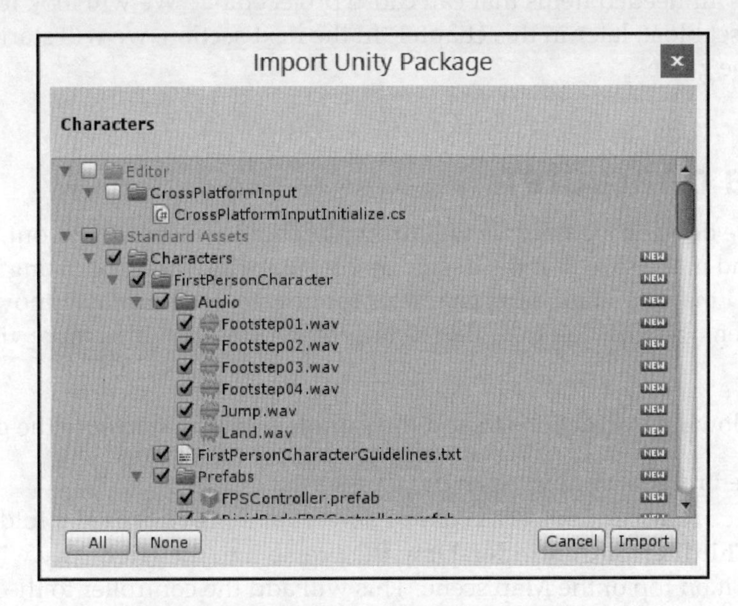

Importing the character's Unity standard asset

5. Note that the **CrossPlatformInput** assets is automatically deselected. The Unity editor recognizes that the project has already imported the standard cross-platform assets. Click on the **Import** button on the dialog to install the **Characters** assets.

6. Finally, let's import the **CrossPlatformInput** assets by selecting the menu item **Assets | Import Package | CrossPlatformInput**. After a few seconds, the **Import Unity Package** dialog will open. Note that the only thing left to import is some fonts. Click on the **Import** button to load the remaining assets into the project.

Good, we should now have all the standard assets we need in the project for the functionality we plan to build in this chapter. Ensure that you open the various new asset folders in the Project window and just familiarize yourself with what new items have been loaded. Assets are a great way to quickly add functionality to your game but can also carry with them many unneeded items that can cause project bloat. We will look for ways to manage that asset bloat, later in this chapter. In the next section, we will start to add those new assets to the game.

Adding a character

Generally, as we develop a game, we will drop placeholder assets to test out game functionality and make sure that the design and vision will work. Following this principle, we will use the Ethan standard asset character for now to get our player movement prototyped. Then, we will look to replace that prototype asset with a more visually pleasing character later.

Perform the following instructions to add the sample Ethan character to the game scene:

1. In the **Project** window, open the `Assets/Standard Assets/Characters/Third Person Character/Prefabs` folder and select the **ThirdPersonController**. Drag the prefab into the **Hierarchy** window and drop it on top of the **Map** scene. This will add the controller to the scene and place the sample Ethan character in the world center, as follows:

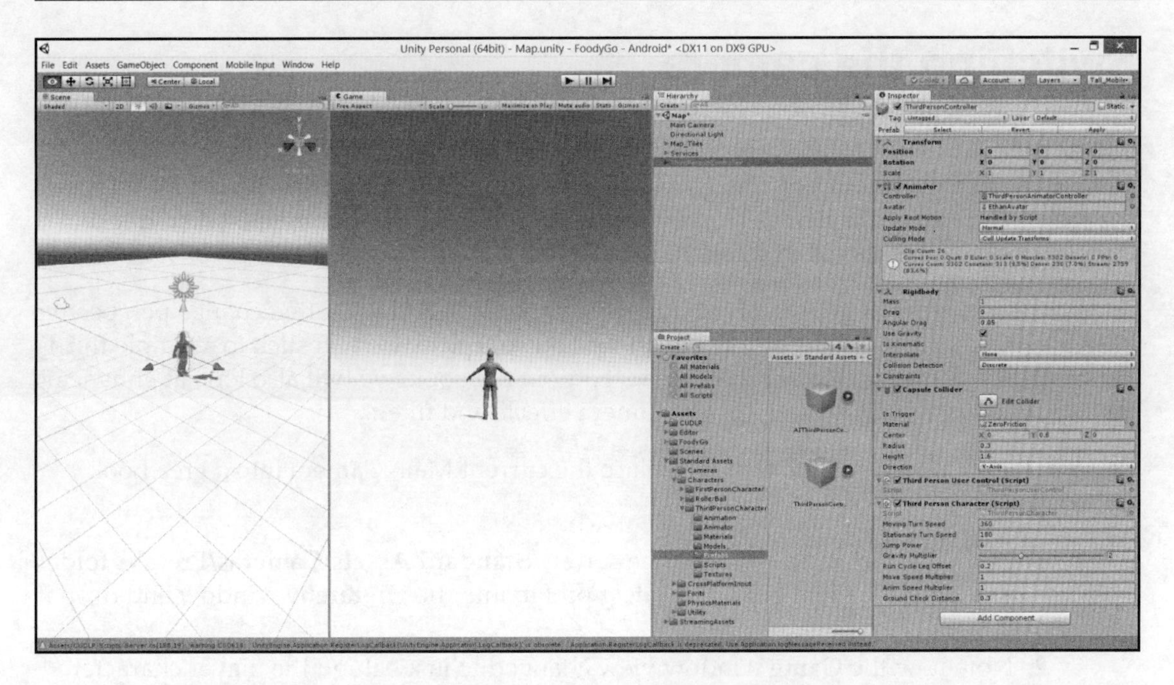

ThirdPersonController loaded in the scene with Ethan character

2. Select the **ThirdPersonController** in the **Hierarchy** window and then rename it to `Player` in the **Inspector** window. This new object will represent the player in the game. It is a convention to name the player object `Player`. Also, many standard scripts will automatically connect to the game object named Player.
3. Hit the Play button to run the game in the editor. Notice, as the game runs, the character becomes animated but is just standing there. If you try to make the character move or jump, it does nothing. Not to worry, this is expected and is because we are using cross-platform input. We will cover input shortly.
4. Press the Play button again to stop the game.

Well, that was certainly easy. We now have a fully rigged and animated 3D character in our scene. Also, the whole process took only a few steps. That is the power of using assets to prototype; but as you will see, we still have quite a bit of work to do. In the next section, we will change the camera our game uses in order to better visualize the player and the game world.

Switching the camera

Probably one of the most critical elements in any game is the camera. The camera is the player's eyes into the virtual game world we create as game developers. In the early days of computer gaming, the camera was generally fixed, but could sometimes pan or move about the scene. Then came the first-person camera and the third-person camera, which tracked the player's movements but in different perspectives.

Today, the game camera has evolved into a cinematic tool and will often change perspective based on the player's actions or movement. For our purpose, we will stick to a simple third-person free look camera for our map scene. Later in the book, we will also look at enhancing the look of our game by adding certain camera effects and filters.

Perform the following instructions to replace the current Main Camera into a Free Look follow camera:

1. In the **Project** window, open the **Assets/Standard Assets/Cameras/Prefabs** folder and drag the **FreeLookCameraRig** prefab into the **Hierarchy** window and drop it onto the **Player** game object.
2. Note how the **Game** window view changed to just behind the player character. That is because the **FreeLookCameraRig** is designed to track our player game object. Look in the **Inspector** window at the **Free Look Cam script** component. You will see a checkbox called **Auto Target Player**. When this option is selected, the script will search the scene for a game object named **Player** and attach itself automatically. The following screenshot shows the **Free Look Cam** component as seen in the Inspector window:

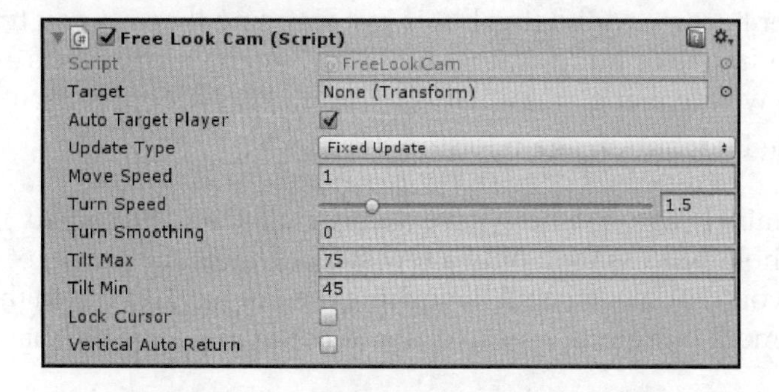

The Free Look Cam component with Auto Target Player enabled

3. Select the **Main Camera** object in the **Hierarchy** window and press *Delete*. This will delete that camera from the scene, as we no longer need it.

4. Press the Play button to start running the game. The game still does not allow any interaction but the view is much better. Let's add some input controls to the game.
5. In the **Project** window, open the **Assets/Standard Assets/CrossPlatformInput/Prefabs** and drag the **DualTouchControls** prefab into the **Hierarchy** window on top of the **Map** scene and drop it. This will add the dual touch control interface as an overlay, which you will see in the **Game** window.
6. From the menu, select **Mobile Input | Enable** to confirm that the mobile input controls are enabled.
7. Press Play to start running the game. You will now be able to move the camera and character around the scene by right-clicking (press *Ctrl* and click on Mac) and holding in the various overlay panels. Here is how your Game window should look now:

Game window with the Dual Touch control interface added

8. Build and deploy the game to your mobile device using the procedure mentioned in `Chapter 2`, *Mapping the Player's Location*. You should now be able to move the camera and character freely around the scene by touching the overlay panels.

Great, with very little effort, we now have a character that can move around the scene with a free look follow camera. Some game developers would be ecstatic at this point. Unfortunately, for us, there are a couple of problems with the current input controls and the way the character moves. We will fix these issues in the next section.

Cross-platform input

Before we get into fixing the input issues, let's understand what cross-platform input is. Cross-platform input creates an abstraction of the input controls, keys, and buttons, which can then be mapped to physical device controls specific to the device when the game is deployed.

For instance, you developed a game for the PC, Mac, and mobile phone. Instead of programmatically checking whether the player fired by left-clicking on a PC, mouse clicked on a Mac or tapping the screen on a phone, you would check whether the player hit the *fire* control. Then, the *fire* control would be defined specific to the device. This would then allow you to easily run your game across platforms or even add additional platforms later. The following is the diagram that shows how this input mapping works:

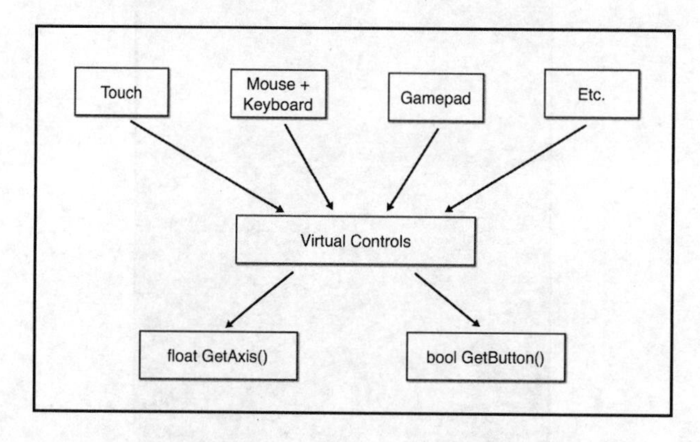

Various input controls mapped into cross-platform input

In the next section, we will show more specific examples of how the cross-platform input functions can be used in scripts.

Fixing the input

As we have mentioned before, the standard assets are great for prototyping functionality, but they have limitations you can quickly bump up against. If you look at the current scene on a mobile device, you will notice a few issues:

- The character in our game should not move from direct player input but rather as a result of the player moving their device. Therefore, we don't need touch input for character movement.
- The jump button is not needed, and we can hide or remove it.
- A player will only be moving the camera or selecting objects and menus by touch, which means we can hide this overlay altogether.

Let's get started fixing these issues by cleaning up the game interface. We want to remove the movement and jump controls and hide the touch pad overlay. The following are the instructions to clean up the movement controls and interface:

1. Select and expand the **DualTouchControls** game object in the **Hierarchy** window. You should see something like this:

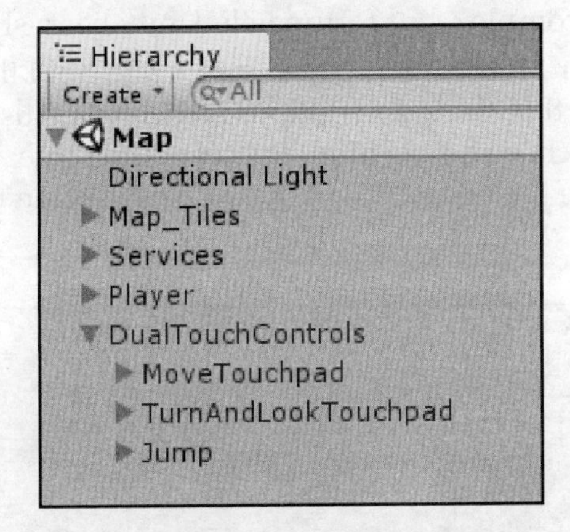

DualTouchControls expanded

2. Select the **MoveTouchpad** game object and press the *Delete* key to delete the object. You will get a prompt informing you that you will break the prefab instance. This is fine, so click on the **Continue** button, as shown in the following dialog:

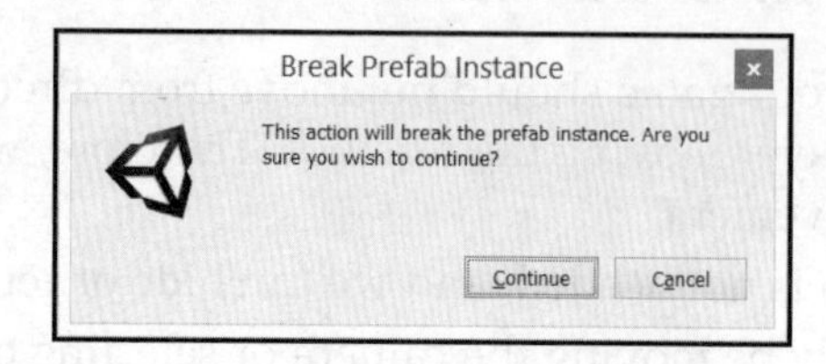

The dialog prompt asking whether you want to break the prefab

3. Select the **Jump** game object and press the *Delete* key to delete the object. Note that both those overlays are now gone in the **Game** window.
4. Select the **TurnAndLookTouchpad** game object. In the **Inspector** window, click on the **Anchor Presets** box to expand the menu. What we want to do is make the **TurnAndLookTouchpad** object fill the entire game screen. That will allow the player to swipe anywhere on the screen to move the camera. The manual process of doing this is complex, but fortunately Unity has a shortcut.
5. With the **Anchor Presets** menu open, press and hold the *Alt* key (the *option* key on a Mac). Note that the menu options switch from just setting the object's anchor to setting the anchor and position. Press and hold the *Alt* key (the *option* key on a Mac) and choose the bottom-right option as shown in the right-hand side of the screenshot:

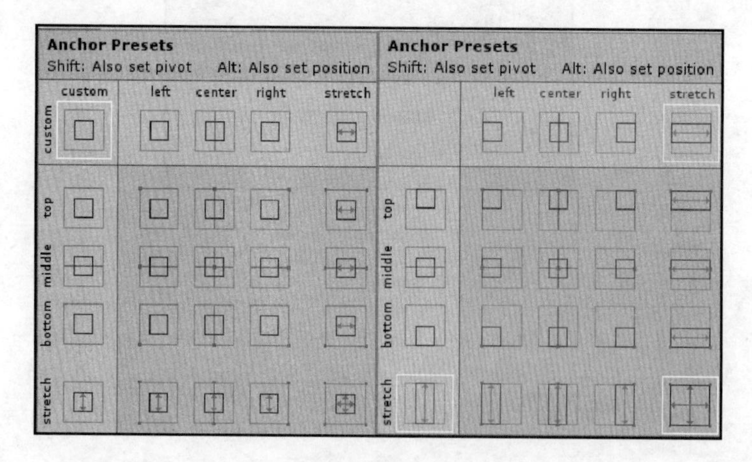

Anchor presets menu, selecting the fill option

6. With the stretch-stretch position or fill option chosen, you should now see the **TurnAndLookTouchpad** fill the **Game** window. Not to worry, if you are confused over what we did and why. We will get into the details of what we just did later when we add the player menus in `Chapter 6`, *Storing the Catch*.
7. Ensure that the **TurnAndLookTouchpad** is selected, and in the Inspector window under the Image component, click on the white box beside the **Color** property. When the color dialog opens, set the **Hex Color** to `#FFFFFF00`. That will make the overlay all but disappear, except for the text.

> Hex color is the hexadecimal number that represents a color. In hex, a color can easily be identified by each of its primary components, namely red, green, blue, and alpha shown as follows: `#RRGGBBAA`.
>
> Each of the component's value range from 00-FF in hexadecimal or 0-255 in decimal.
>
> Alpha represents the opacity, with FF being fully opaque and 00 being fully transparent.
>
> `#FF0000FF` would be red, `#000000FF` would be black, and `#FFFFFF00` would be transparent white.

8. Select and expand the **TurnAndLookTouchpad**. Select the **Text** object and press the *Delete* key to delete the object.
9. Run the game in the editor and deploy it to your device. Notice that now only the camera is controlled by swiping.

Great, cleaning up of the movement controls and interface helped us fix a couple of items on our list. Now, in order to fix the player character's movement, we will need to create our own controller scripts. Unlike most games, we actually don't want the player to directly control their character. Instead, the player needs to physically move their device in order to move around the virtual world. Unfortunately, that also means we will need to modify some of the previous scripts we already imported and set up. That really is just the process of development and especially game development. We will attempt to minimize rewrites as we move through this book, but it is important you understand that it is part of the development process.

We will start by creating a new compass and GPS controller script. This script will move the player on the map by tracking the device's GPS and compass. Perform the following instructions to create this script:

1. Expand the `Assets/FoodyGO/Scripts` folder in the **Project** window. Select the **Scripts** folder and create a new folder by selecting menu item **Assets** | **Create** | **Folder**. Rename this folder `Controllers`.
2. Select the new **Controllers** folder and select menu item **Assets** | **Create** | **C# Script** to create a new script. Rename this script `CharacterGPSCompassController`; It's a verbose name but also descriptive.
3. Double-click on the new script to open it in MonoDevelop or the editor of your choice. You should see the following default code listing, for the new script:

```
using UnityEngine;
using System.Collections;

public class CharacterGPSCompassController : MonoBehaviour {
  // Use this for initialization
  void Start () {
  }
  // Update is called once per frame
  void Update () {
  }
}
```

4. We will start simple and just handle the compass part of the controller. Using the device's compass, we will be able to always orient the player in the direction the device is pointing, while they are not moving. When the player moves, we will always orient the player toward the direction of travel. Add the following line of code inside the `Start()` method:

```
Input.compass.enabled = true;
```

5. The `Start()` method is for initialization, and that line of code essentially turns the device's compass on for reading a heading. Now that the compass is on, add the following code to acquire the heading in the `Update()` method, as follows:

```
void Update()
    {
        // Orient an object to point to magnetic north and adjust for
map reversal
        var heading = 180 + Input.compass.magneticHeading;

        var rotation = Quaternion.AngleAxis(heading, Vector3.up);
```

```
        transform.rotation = rotation;
    }
```

6. The first line of code in the `Update()` is a comment, telling us why and what the next couple of lines are for. When we write comments, we don't just want to explain what we are doing but also why we are doing it. Often, the reason why will be more important. Get into the habit of writing comments in your code. Commenting is never a wasted effort.
7. The next line of code sets a variable called `heading`, with the value of the compass magnetic north offset by 180. We add 180 degrees to the compass reading in order to orient the character to align with the tile maps north. The tile map, as you may recall, is reversed in order to simplfy the math.
8. The following line may look strange, especially if you don't know what a quaternion is. A quaternion is a number system that extends the space of complex numbers. This sounds like quite a mouthful, especially if you, as of yet, don't possess an advanced level of math. Without getting too far off topic, for now, think of a quaternion as a helper that allows us to easily define a rotation anywhere in 3D space. This means the call `Quaternion.AngleAxis(heading, Vector3.up)` is defining a rotation around the world up or *y* axis. That value is being stored in a local variable called `rotation`. The following is a diagram you can use to remember what each axis and corresponding vector is called using the left-handed coordinate system Unity uses:

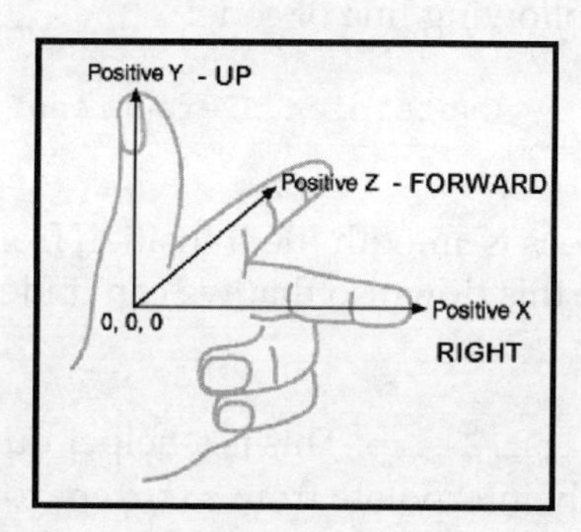

Left-handed coordinate system explained

9. The last line of code sets the `transform.rotation` to the value of `rotation` calculated with a quaternion helper. After you enter the last line of code, ensure that you save the file in MonoDevelop or the code editor of your choice. Then, go back to Unity and wait for a couple of seconds while the new script gets compiled into the game. Unity has a great auto compilation feature that will recompile the entire project whenever a file has been changed.

10. Select the **CharacterGPSCompassController** script in the **Project** window. Then, drag and drop the file onto the **Player** game object. This will add the script to the `Player`.
11. Select the **Player** game object in the **Hierarchy** window. In the **Inspector** window, select the gear icon beside the **Third Person User Control (Script)** component. From the context menu, select **Remove Component** to remove the script from the `Player` since we no longer want to control our character using the standard input script.
12. Build and deploy the game to your mobile device to test. Just like when we tested the GPS functionality, the compass will only return values when running on an actual device with a compass.
13. Test the game by moving the device in various orientations. Notice how the character moves as the device rotates. One thing you may notice, however, is that movement is not fluid, and it is quite jumpy or bouncy. The jitter is caused by constantly updating the player with every new reading from the compass. If you have ever looked at a physical compass, you will see exactly the same thing. Of course, we don't want our character twitching, so let's take a look at how we can fix that.
14. Open back up the **CharacterGPSCompassController** script in the `MonoDevelop` editor by double-clicking on it in the **Project** window.
15. Change the last line in the `Update()` method from `transform.rotation = rotation;` to the following line of code:

```
    transform.rotation = Quaternion.Slerp(transform.rotation, rotation,
Time.fixedTime*.001f);
```

16. What this change does is smooth the transition from one heading change to another. Let's break this down so that we can understand the details of what is happening:

    - `Quaternion.Slerp`: This is a helper quaternion function to spherically interpolate from a current rotation into the new rotation. Don't get confused by the wording spherically interpolate. Essentially, all that it means is we are smoothing the rotation across a sphere by adding additional smoothing points. Lerp is linear interpolation between two data points. In the following diagram, $\mathbf{q_a}$ is the start, $\mathbf{q_b}$ is the end, and $\mathbf{q_{int}}$ represents a calculated smoothing point:

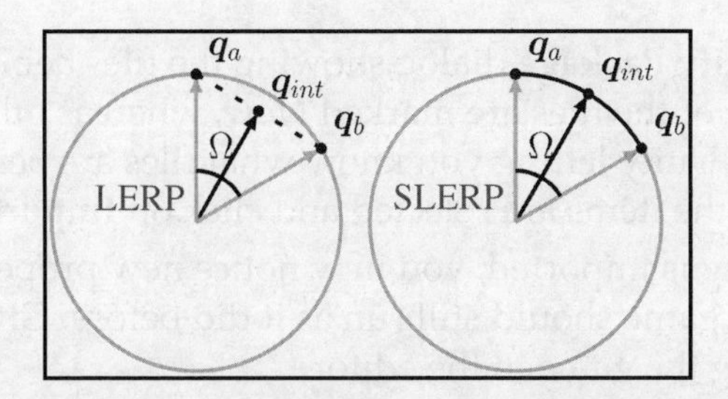

Shows how smoothing points are calculated when using LERP or SLERP

- `transform.rotation`: This indicates the object's current rotation.
- `rotation`: This is the rotation we want to change to, that is, the value calculated in the preceding line.
- `Time.fixedTime * .001f`: This describes how much we want the rotation to change in one call. `Time.fixedTime` is the amount of time one frame of the game should render in. We multiply that number by `.001f` in order to make the rotation change very small during each frame. Feel free to change this value to see the effect it has on smoothing the rotation.

17. When you are done with editing the script, save the file. Then, go back to Unity and then wait for a few seconds for the scripts to recompile.
18. Build and deploy the game again. Test the game and notice the difference in how smooth the character now slowly turns.

Good, we fixed a number of the interface issues and now have the character turning relative to the device's orientation. We accomplished that by modifying the standard assets we imported and writing a simple controller script. Unfortunately, we still have a lot of scripts to add and rewrite in order to get our character to walk or run over the map. Since we don't really have time-and I doubt you have the patience-to write and review the script changes line by line, we will instead import all the updated scripts and then review the important sections in more detail.

Let's import the updated scripts by performing the following steps:

1. From the Assets menu, select **Import Package | Custom Package...**
2. After the **Import package** dialog opens, navigate to the downloaded code folder `Chapter_3_Assets` and open **Chapter3.unitypackage** by selecting the file and clicking on **Open**.

3. The **Import Unity Package** dialog showing the files being imported will open. Note that some of the files are marked **New**, whereas others have a refresh symbol. This is Unity letting you know what files are being changed or added. Ensure that all the items are selected and click on **Import**.
4. After everything is imported, you may notice new properties on a few game objects, but the game should still run as it did before. Give it a try by pressing Play and testing the game in the editor.

So, for the most part, those updated scripts will still run our game as expected. However, we still don't see our character moving around the map. In order to get our character moving, we will need to set some new properties on the scripts. However, before doing that, we should understand what was changed. Each of the following sections review a script component and the changes that have been done.

Here is a summary and review of the scripts we imported and updated:

- **Controllers**:
    - `CharacterGPSCompassController`: This script was updated to consume GPS readings from the GPS Location service.
- **Mapping**:
    - `Geometry`: This file is designated for custom spatial types. A new type called the `MapEnvelope` was added.
    - `GoogleMapTile`: This script is almost identical to our previous version; only a couple of lines were added.
    - `GoogleMapUtils`: This is our library of spatial math functions. A couple of new functions were added in order to convert between map and game world scale.
- **Services**:
    - `GPSLocationService`: Numerous code changes were needed to support a new map drawing strategy. A way to simulate GPS readings was added to aide testing and development.

GPS location service

The first script we will look at is the **GPS Location Service**. Expand the **Services** object in the **Hierarchy** window and select the **GPS** service object. Review the object in the **Inspector** window and notice all the new properties and sections. Two new sections have been added to the GPS Location Service script. The first section is for map tile parameters and the second section is a new feature for simulating GPS data. We will review the purpose and properties of these two subsequent new sections.

Map tile parameters

Previously, whenever our GPS service acquired a new update from the device, that data would automatically be pushed to the map and the map would redraw itself. As you saw, that simple method worked, but it did suffer from a couple of problems. First, every time the service acquired a new location, the map would make several expensive calls to refresh itself. It didn't matter whether the device moved a meter or 100 meters. Second, if we want to show our character move across the map, then we can't refresh the map each time with every new location. Instead, we only want to refresh when the character reaches a tile boundary. Fortunately, we can solve both of these issues by allowing the GPS service to track the size of a map tile and then call a map refresh when a new GPS reading is outside the current center tile.

The following is a diagram showing how this will work:

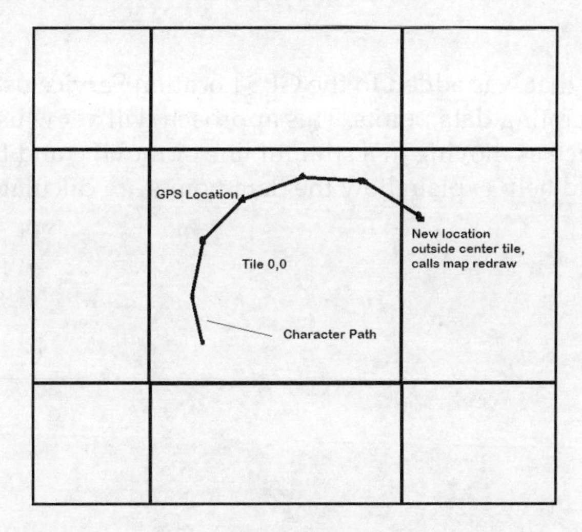

GPS tracked on map tile boundary

In order for the GPS Location Service to be able to keep track of the tile boundary, it needs to know how the tile is generated. This is why we need to pass the same parameters we used to construct the map tiles into the service. Here is a review of those parameters and what they should be set to for the game:

- **Map tile scale**: This indicates the scale of the map tiles. We set this to 30 for the current map.
- **Map tile size pixels**: This is the map tile image size being requested from Google. This was set to 640 pixels.
- **Map tile zoom level**: This is the zoom level or the scale of the map. We chose a value for 17 in order to neighborhood scale map.

The Google Static Maps API throttles the number of requests an IP address or a device can make every 5 seconds and every 24 hours. Currently, the limit is 1000 requests per 24 hours.

GPS simulation settings

As you have probably already realized, testing the GPS service was going to be difficult. Sure, we set up CUDLR to allow us to view real-time updates while the application was running on the device, but that had limitations. Ideally, we want to be able to test the way our game objects consume the GPS service as it is running in the Unity editor. That way, we can see how the game will run, without having to move around the house or office every few seconds. We can accomplish this testing by generating simulated location readings from the GPS service.

The simulation service that was added to the GPS Location Service uses a simple origin offset approach to generating data points. This approach will allow us to define simple movement patterns, such as moving in a straight line or moving and turning. The following is a diagram that should help explain how the data points are calculated:

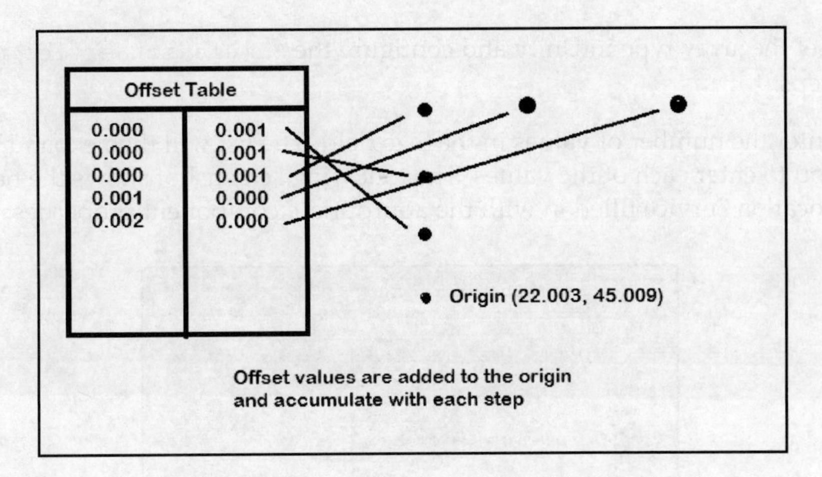

Offset data approach to simulate GPS readings

In the subsequent list, each of the GPS simulation properties are explained in more detail:

- **Simulating**: If checked, the GPS service will generate simulated data. Data simulation will not run on a mobile device. Check this option to turn on the GPS simulation.
- **Start coordinates**: This is the origin point for the simulation. Offset values will be accumulated against the origin. Use our test coordinates or some latitude or longitude coordinates that are more familiar to you.
- **Rate**: This is the rate in seconds at which new GPS readings will be simulated. A good value for this is 5 seconds.
- **Simulation offsets**: This is the offset table of array values that will be added and accumulated against the origin. These values are in latitude or longitude values, so keep these numbers small. A good starting value is around +- 0.0003. The offset values will continually loop. So, after the bottom value is added, the offset table will start at the top again.

In order to set the array type in Unity and configure the `SimulationService`, perform the following steps:

1. Enter the number of values in the Size field. The list will then expand to allow you to enter each of the values. The following screenshot shows the new GPS Location Service filled in with the appropriate component properties:

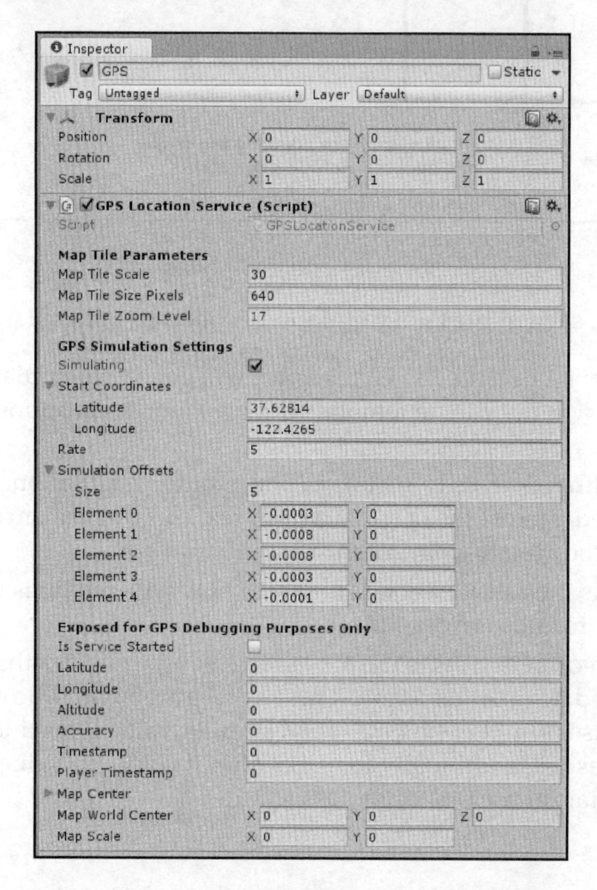

Inspector view of GPS service object

2. Select the **Player** game object in the **Hierarchy** window. In the Inspector window, you will note that a new property was added to the **Character GPS Compass Controller** script component. We now need to set the GPS Location Service in this script, just like we did for the map tiles. That makes sense, because the character controller also needs to consume updates from the GPS service. Here is a screenshot of the controller script:

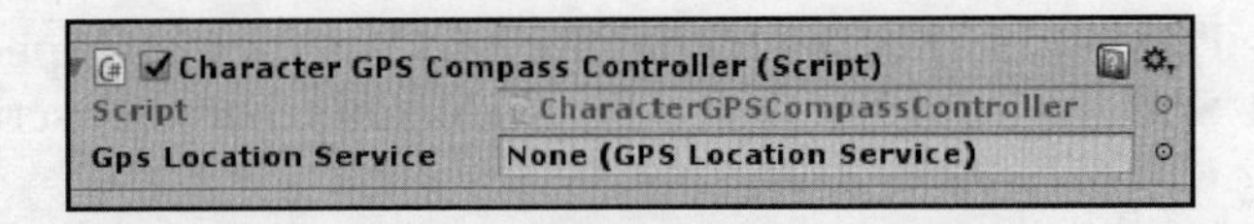

Character GPS Compass Controller script

3. With the **Player** object still selected, drag the **GPS** object from the **Hierarchy** window and drop it into the **Gps Location Service** field on the **Character GPS Compass Controller** component.
4. Run the game in the editor by pressing the Play button.

Now, as the GPS data is being simulated, you will see the character move around the map. However, as you probably noticed, we now have a couple of other issues. The camera no longer remains at a fixed distance from the character, and the movement is too quick. Fortunately, these are simple fixes and can be done quickly by following the directions, as follows:

1. In the **Hierarchy** window, expand the **Player** object and select the **FreeLookCameraRig**. The reason the camera is not following the `Player` is because it is tracking the game object but not the game object's transform. This is a subtle but important difference. We will need to set the target transform in the camera to be the `Player`. Here is a screenshot that shows the empty transform field in the **Free Look Cam** script component:

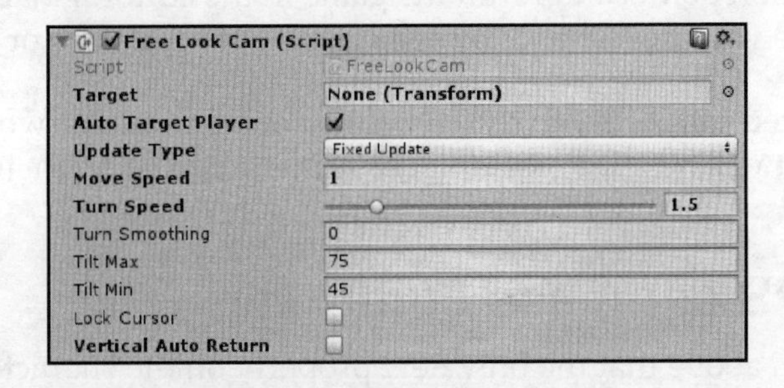

Free Look Cam script properties

2. While the **FreeLookCameraRig** object is still selected in the **Hierarchy** window, drag the **Player** object to the Target field in the **Free Look Cam** script component. This will now force the camera to follow the character's transform.

3. Select the **Player** object in the **Hierarchy** window. In the **Inspector** window, set the **Move Speed Multiplier** of the **Third Person Character** script component to **0.1**, as shown in the next screenshot:

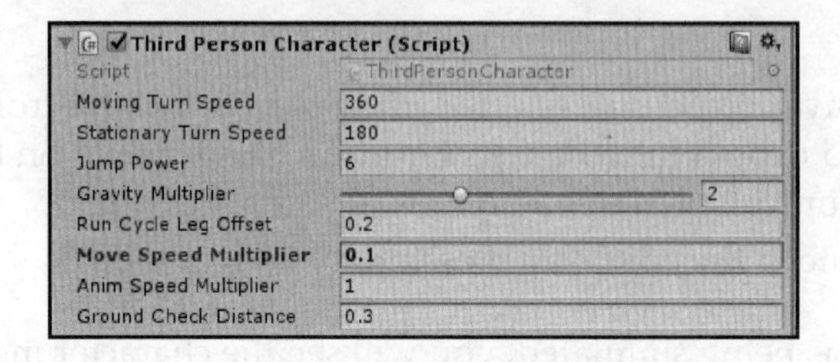

The Third Person Character script component

4. The reason we set this to a small value is to account for our difference in map scale. The typical character controller will move a character at a speed close to walking. This is OK, when your game is at a 1:1 scale. Our game, however, is at a much grander scale. Calculating what this exact value is will depend on a number of factors. For now, we will estimate a value of 0.1 to reduce the character speed by one-tenth. We will add this as item we will discuss in `Chapter 9`, *Finishing the Game*.
5. Press the Play button to run the game in the editor. As you can see, the character moves around the map as expected. Ensure that you try other offset values for the simulation points and rerun the game in the editor. Finally, build and deploy the game to your device and take the game out for a walk or drive.

This covers the updated properties to the GPS Location Service. We won't cover the specific script changes, as that will be left to the diligent reader to do so on their own.

Character GPS compass controller

If you remember from above that the only new property on the character compass GPS controller was a reference to the GPS Location Service. This was needed in order to update the player's position as new GPS readings are acquired. Since we began writing the character controller script from the beginning, let's review the code to see what has changed.

Locate the Character GPS Compass Controller script in the Project window under the Assets/FoodyGO/Scripts/Controllers folder. Double-click on the script to open it in MonoDevelop or the editor of your choice. After the script opens, take a few minutes to review the changes.

Now, let's go through each section of the script and understand the changes in more detail. Starting from the top, following are the top lines of the script:

```
using UnityEngine;
using UnityStandardAssets.Characters.ThirdPerson;
using packt.FoodyGO.Mapping;
using packt.FoodyGO.Services;

namespace packt.FoodyGO.Controllers
{
    public class CharacterGPSCompassController : MonoBehaviour
    {
        public GPSLocationService gpsLocationService;
        private double lastTimestamp;
        private ThirdPersonCharacter thirdPersonCharacter;
        private Vector3 target;
```

At the very top is the using, which is standard for a C# script. The top line is standard for all Unity scripts. The next few lines import the other types this script will use. Then, the start of the script definition begins with a namespace declaration. Defining a namespace is standard when writing a C# file in other development platforms, but it is a practice Unity does not require or enforce. Unity does this in order to support a variety of scripting languages. However, as you may painfully learn, not adhering to a namespace can cause all manner of naming conflicts. For the purpose of this book, we will adhere to the `packt.FoodyGO` namespace.

After the namespace declaration is the class definition followed by some new variables. A variable for the GPS Location Service was added and made public so that it could be changed in the Unity editor. Then, three new private variables were added. We will review the purpose for each of those variables below in the relevant code section.

Marking a variable marked private sets the variable for internal use only within the class. If you are an experienced C# developer, you may also be asking, why we are using a public variable when it should be a property accessor. A public variable can be modified in the Unity editor, but a private or property accessor is hidden. You can, of course, still use property accessors from other types, but in general, most Unity developers will avoid them, instead prefer to just use public or private.

Here is an example of a property accessor:

```
public double Timestamp
        {
            get
            {
                return timestamp;
```

```
            }
            set
            {
                timestamp = value;
            }
        }
```

The next section of the script we will look at is the `Start` method shown in the code listing:

```
    // Use this for initialization
        void Start()
        {
            Input.compass.enabled = true;
            thirdPersonCharacter = GetComponent<ThirdPersonCharacter>();
            if (gpsLocationService != null)
            {
                gpsLocationService.OnMapRedraw +=
GpsLocationService_OnMapRedraw;
            }
        }
```

As you can see, we added a couple more lines after the `Input.compass.enabled` line we wrote earlier in the chapter. After turning on the compass, the next line gets a reference to the `ThirdPersonCharacter` component script and stores it into that private variable `thirdPersonCharacter`. The `ThirdPersonCharacter` script controls the movement and animations of our character. As you will see, we will use that reference to move the character later in the `Update` method.

The following line checks whether the `gpsLocationService` is not null. If the value is not null, as it shouldn't be, we consume a new event on the GPS service called `OnMapRedraw`. `OnMapRedraw` fires after the center map tile is recentered and redrawn. Remember from our explanation above that the GPS service now tracks when it needs to recenter the map. After the service initiates a redraw to the map tile, the map tile requests a new map image. After the image request returns and the map tile is changed, the map tile tells the GPS service that it has refreshed itself. The GPS Service will then broadcast the `OnMapRedraw` event to all its consumers to let them know that they also need to recenter themselves. If you are a little lost on how this all connects, hopefully the following diagram will help:

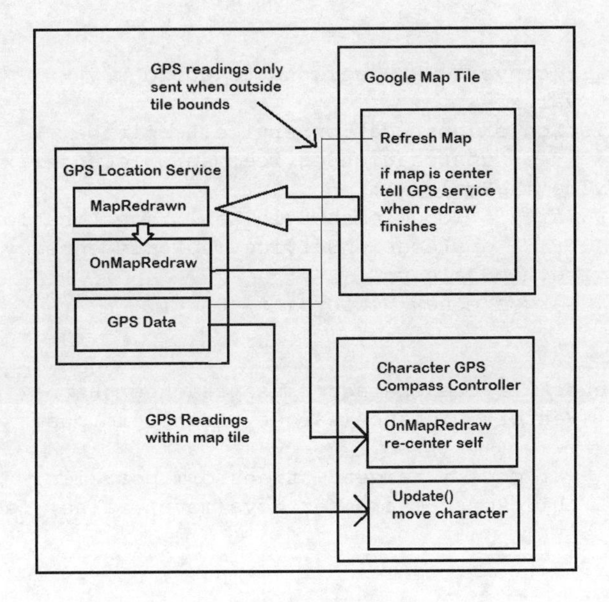

Flow of data and events from GPS service

The next couple of lines after the Start method are for subscribing to the `OnMapRedraw` event. The event is a void method that passes the event source in the `GameObject g`, parameter:

```
private void GpsLocationService_OnMapRedraw(GameObject g)
    {
        transform.position = Vector3.zero;
        target = Vector3.zero;
    }
```

When the `OnMapRedraw` fires, it tells the character controller it needs to reset its position back to the origin, as the map has finished redrawing. Inside the event handler, the `Player` transform position is to set to `Vector3.zero`; this is the same as setting the position to (0,0,0). Likewise, we do the same to the `target` variable. We will get to that variable shortly in the `Update` method.

Finally, we come to the last method and the real worker in our class, the `Update` method. The following is the code listing:

```
// Update is called once per frame
    void Update()
    {
        if (gpsLocationService != null &&
            gpsLocationService.IsServiceStarted &&
            gpsLocationService.PlayerTimestamp > lastTimestamp)
```

```
            {
                //convert GPS lat/long to world x/y
                var x =
((GoogleMapUtils.LonToX(gpsLocationService.Longitude)
                    - gpsLocationService.mapWorldCenter.x) *
gpsLocationService.mapScale.x);
                var y = (GoogleMapUtils.LatToY(gpsLocationService.Latitude)
                    - gpsLocationService.mapWorldCenter.y) *
gpsLocationService.mapScale.y;
                target = new Vector3(-x, 0, y);
            }

            //check if the character has reached the new point
            if (Vector3.Distance(target, transform.position) > .025f)
            {
                var move = target - transform.position;
                thirdPersonCharacter.Move(move, false, false);
            }
            else
            {
                //stop moving
                thirdPersonCharacter.Move(Vector3.zero, false, false);

                // Orient an object to point to magnetic north and adjust
for map reversal
                var heading = 180 + Input.compass.magneticHeading;
                var rotation = Quaternion.AngleAxis(heading, Vector3.up);
                transform.rotation = Quaternion.Slerp(transform.rotation,
rotation, Time.fixedTime * .001f);
            }
        }
```

As you can see, a number of lines of code were added to support the GPS movement of our character. The code may seem complex, but it is fairly straightforward if we take our time.

At the top of the method, you will see almost the same test we did in the Google Map Tile script to check whether the GPS service is set, is running, and sending new location data. Inside the `if` statement is a couple of complex calculations that use the `GoogleMapUtils` helper library to convert GPS latitude or longitude to `x` or `y` 2D world coordinates. This is then converted into 3D world coordinates in the next line and stored in that target variable. Ensure that you notice the negation on the `x` parameter. Remember that our map is flipped and positive `x` points to west and not east. The target variable stores the 3D world position we want our character to move to.

The next `if/else` statement checks whether the player has reached the target position. Normally, this test is done with a value of `0.1f`. However, that is at a real-world 1:1 scale; again, for our purpose, we will use a much smaller number.

Inside the `if` statement, we know the character has not reached its destination and needs to continue moving. In order to make the character move, we need to pass the `thirdPersonCharacter` a move vector. The move vector is calculated by subtracting the character's current position, denoted by `transform.position`, from the `target`. The result is a vector that we then use to call the Move method on the `thirdPersonCharacter`. Internally, the `ThirdPersonCharacter` script will manage the animations and movement.

In the else section of the`if` statement, we know the character is not moving or at least shouldn't be. Therefore, we call the `Move` method on the `thirdPersonCharacter` again, this time with a zero vector in order to stop movement. After that, we check the compass heading and set that just as we did before. Note that we only set the compass heading if the character is not moving. After all, when the character moves, we want them facing in the direction of travel.

Well, that completes reviewing the `CharacterGPSController` script. This script is a good start for showing player movement around a map. However, as you play the game or allow others to play the game, you may notice some areas that may need improvement. Feel free to improve on this script as much as you like and make it your own.

Swapping out the character

Now that we have everything working as expected, let's take some time to visually improve our player's character. We certainly don't want our gray Ethan character to ship with our game. Of course, this game is about game development and not 3D modeling, so we will want to use something easily available. If you open up the Unity Asset Store and do a search for 3D characters, you will see plenty of assets available. Refine this search to just free assets and there is still a lot. So, what is the best option? When it comes down to it, the best option is what works for you and your team. Feel free to try other character assets.

For this book, we will use the base `iClone` characters, which are freely available from the asset store. This is an excellent resource and they deserve the five-star rating they had at the time of writing this book. The asset packages are concise and don't have any unnecessary content, a big plus for mobile games. Also, the character models are low in polygon count, which is import to mobile rendering.

Perform the following directions to import an `iClone` character and swap it for the Ethan character:

1. Start by opening the **Asset Store** window by selecting menu item **Window | Asset Store**.
2. After the window opens, enter `iclone` in the search box and press *Enter* or the search icon.
3. When the search completes, at the top of the list should be the three base iClone characters **Max**, **Izzy,** and **Winston**, as follows:

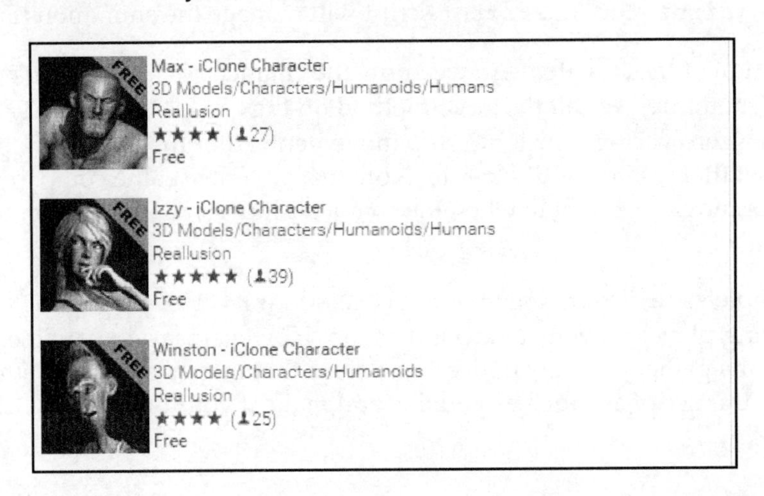

Base iClone character options

4. Now, at this point, you can choose the character you want to use for your version of FoodyGO. All of the characters will function the same, and there is only one minor setup difference. Feel free to select one character now and then come back later to add a different base character. Select a character to continue; this choice is entirely up to you.
5. The characters asset listing will load in the window, and on the listing will be a button to download and import the asset. Click on the **Download** button. It may take a few minutes to download, so you know the drill, grab a coffee or your favorite beverage and wait for it to finish.
6. After the asset is downloaded, you will see a **Import Unity Package** dialog prompting you to select what you want to import. Ensure that everything is selected and click on the **Import** button.
7. When the import completes, you will see that a new folder is added to the **Assets** folder in the **Project** window. This new folder will be named according to your character selection, either `Max`, `Izzy`, or `Winston`.

8. Expand the character-named folder and select the **Prefab** folder. You will see a prefab with your character's name on it. Select and drag the prefab from the **Project** window and drop it onto the **Player** object in the **Hierarchy** window.
9. Select and expand the **Player** object in the **Hierarchy** window. Ensure that your new character is added to the **Player** object. Select the character object and reset the transform. You can reset the transform by clicking on the gear icon in the **Transform** component from the **Inspector** window. Then, select the **Reset** option from the context menu.

You should now see your character overlapping the Ethan character, as follows:

iClone character overlapping the Ethan character

The 3D character is designed by Reallusion iClone Character Creator. To create more customized characters, please visit `http://www.reallusion.com/iclone/character-creator/default.html` for more details.

1. While in the **Inspector** window, disable the **Animator** component by unchecking the checkbox beside the text **Animator**. Look at the **Avatar** field in the animator and remember the name that is populated there. This will be different for each character, so write it down if you need to.

2. Go back to the **Hierarchy** window and select the **EthanBody** object under the **Player** and press the *delete* key to remove the object. Also, delete the **EthanGlasses** and **EthanSkeleton**.
3. Select the **Player** object. Change the **Avatar** property of the **Animator** by clicking on the bullseye icon in the **Inspector** window. A **Select Avatar** dialog consisting of several names will open. Choose the name that matches what you wrote down in step 10. Then, close the dialog.
4. Run the game in the editor by pressing the Play button. Your new character should be animating and moving if you run the GPS service in simulation mode. Ensure that you build and deploy the game to your mobile device.

As you can see, the process is pretty simple to quickly switch out a character, so try any number or all of the iClone characters. If you have another character asset in mind, you could also try that as well. Of course, the possibilities are endless. Here is an example of the three different iClone characters in our game:

Three different base iClone characters in game.

Summary

Well, this certainly had to be our busiest chapter yet, and we covered several items. First, we imported the standard assets for characters, cameras, and cross-platform input. Then, we added a player character into our map scene with a camera and touch input controls. We then wrote a new script for controlling the player character with the device's compass and GPS. After that, we determined that our GPS service needed a simulation mode and a way to track GPS readings over the map. We then imported a number of updated scripts and configured those to correctly move and animate the game character around the map. Finally, we decided the standard asset character was too bland for our tastes, and we spiced up the game by importing and configuring an iClone character.

In the next chapter, we will continue to explore gameplay and allow the player to interact with world objects. We will spawn the *catch* on the map and allow the player to track the creatures. This will require us to do more script writing, UI development, custom animation, and some special effects.

4

Spawning the Catch

Now that we have our player moving around the real and virtual world, it is time to get into other aspects of our game. If you recall, in Foody GO, players need to catch the foody monsters. Foody monsters are genetically engineered lab accidents. These monsters now wander everywhere and also happen to be amazing cooks and chefs. After a monster is caught, the player can train it to be a better cook or take it to restaurants to work and earn points. With that bit of background established, in this chapter we will work toward spawning and tracking the monster creatures around the player.

This chapter will be a mix of working with Unity and writing or editing new scripts. The design of our game is unique enough that we cannot rely on standard assets anymore. Also, we have previously avoided the complexities of GIS and GPS math in order to not bog us down. Instead, we opted to only briefly mention the GIS library functions. Fortunately, you should have enough basic GIS knowledge now to take your learning to the next level and get into the math. If math isn't your subject, that is fine, as we will also be covering the following topics:

- Creating a new monster service
- Understanding distance in mapping
- GPS Accuracy
- Checking for Monsters
- Projecting coordinates to a 3D world space
- Adding a Monster to the Map
- Building the monster prefab
- Tracking monsters in the UI

Before we get into it, if you have Unity from the last chapter opened, with the game project loaded, then move on to the next section. Otherwise, open up Unity and load the FoodyGO game project or open the `Chapter_4_Start` folder from the downloaded source code. Then, ensure that the Map scene is loaded.

When you open up one of the saved project files, you will also likely need to load the starting scene. Unity will often create a new default scene rather than trying to guess which scene should be loaded.

Creating a new monster service

Since we need a way to track the monsters around the player, what better way to do this than with a new service? The monster service will need to accomplish a few jobs, as follows:

- Track the players location
- Query for monsters in the vicinity
- Track monsters within range of the player
- Instantiate a monster when it is close enough to the player

For now, our monster service will only query and track monsters local to the player's device. We are not creating a monster web service where multiple players will consume and see the same monsters, yet. However, in `Chapter 7`, *Creating the AR World*, we will convert our service to use an external service to better populate our monsters. Open up Unity and follow these directions to start writing the new service script:

1. In the **Project** window, open the `Assets/FoodyGo/Scripts/Services` folder. Right-click (press *Ctrl* and right-click on a Mac) to open the context menu and select **Create** | **C# Script** to create a new script. Rename the script `MonsterService`.
2. Double-click on the new `MonsterService.cs` file to open it in `MonoDevelop` or the code editor of your choice.
3. Just after the usings and before the class definition, add the following line:

```
namespace packt.FoodyGO.Services {
```

4. Scroll down to the end of the code and finish the namespace by adding an ending:

```
}
```

5. As you may recall, we added a namespace to our code in order to name conflicts and for organization.
6. Just inside the class the definition, add a new line:

```
public GPSLocationService gpsLocationService;
```

7. That line allows us to add a reference to our GPS service in the editor. Remember, we want to track the player's location also. After you are done with editing, save your file.
8. Confirm that your script now looks like the following:

```
using UnityEngine;
using System.Collections;

namespace packt.FoodyGO.Services
{
    public class MonsterService : MonoBehaviour
    {
        public GPSLocationService gpsLocationService;
        // Use this for initialization
        void Start()
        {

        }

        // Update is called once per frame
        void Update()
        {

        }
    }
}
```

Now, we will add the new script to a new **MonsterService** game object in the **Hierarchy** window using the following steps:

1. Go back to Unity and make sure that the Map scene is loaded.
2. Select and expand the **Services** object in the **Hierarchy** window.
3. Right-click (press *control* and right-click on Mac) on the **Services** object, to open the context menu and select **Create Empty** to create a new empty child object.
4. Select and rename the new object **Monster** in the **Inspector** window.
5. Open the `Assets/FoodyGo/Scripts/Services` folder in the **Project** window.
6. Drag the new `MonsterService` script and drop it onto the new **Monster** object.
7. Select the **Monster** object in the **Hierarchy** window.

8. Drag the GPS object, also under the **Services** object, to the **Gps Location Service** slot for the `Monster Service` script component on the **Monster** object in the **Inspector** window. The following screenshot shows what this will look like:

Monster service added and configured

Obviously, we have a number of things left to do, but this is a good first step in adding our new Monster Service. In the next section, we will get into the math behind calculating distances and world position. After that, we will get back to adding more functionalities to our new service.

Understanding distance in mapping

In the previous chapters, we didn't have to worry about distance because we were only working with a fixed point, that is, the player's position. Now, we want the player to search out and find monsters. We want our Monster service to determine whether a monster is close enough to be seen or heard. In order to do that, our Monster service needs to be able to calculate distance between two mapping coordinates: the player's position and the monster's hiding location.

You may ask, "Why is that so difficult, doesn't Unity do that all the time?" The answer is yes and no. Unity is great at calculating linear distances between two points in 2D and 3D space. However, remember that our mapping coordinates are actually in decimal degrees around a sphere, the earth. In order to properly calculate the distance between two points on a sphere, we need to draw a line on the sphere and then measure the distance. Here is a diagram that hopefully should explain that better:

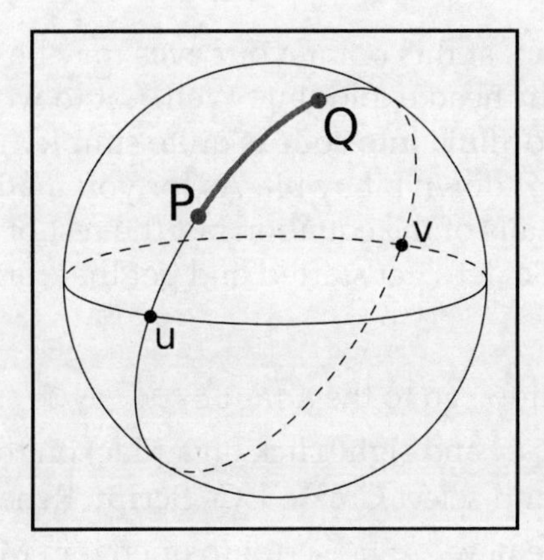

Measuring the distance between two points on a sphere

As the preceding diagram shows, we are measuring the arc and not the straight line distance between coordinate points **P** and **Q**. As an exercise, think about measuring the distance between **u** and **v** in the diagram. Imagine you were flying around the world from city at point **u** to city at point **v**. Which method would you hope your airline used to calculate the fuel?

In order to correctly find the distance between two map coordinate locations, we use a formula called the haversine, as shown in the following diagram:

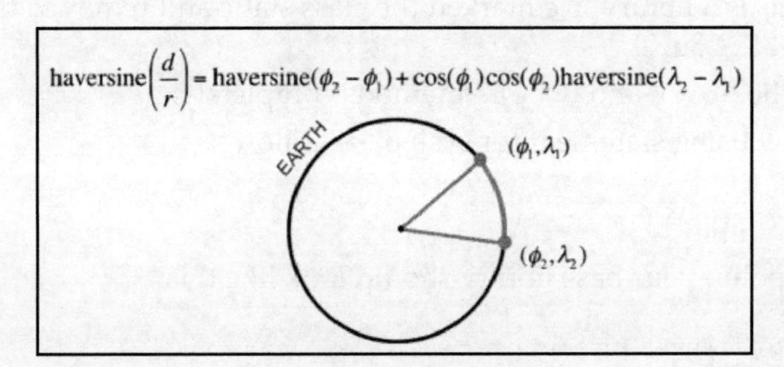

After a little algebraic manipulation, the preceding formula reduces to the form:

$$d = 2r \arcsin\left(\sqrt{\sin^2\left(\frac{\phi_2 - \phi_1}{2}\right) + \cos(\phi_1)\cos(\phi_2)\sin^2\left(\frac{\lambda_2 - \lambda_1}{2}\right)}\right)$$

haversine formula for distance

If you are not a mathematician, at this point, your eyes may be closing or maybe struggling to continue on, for fear of your head exploding. Well, not to worry. As intimidating as those equations look, getting that formula into code is quite simple. The trick is to not get overwhelmed and just break it down bit by bit. As for you mathematicians, apologies for not going further into the details of the equation but the rest of us are sure that you can work that out on your own. So, let's get started and get that formula into a new `Math` library:

1. In the **Project** window, go to the `Assets/FoodyGo/Scripts/Mapping` folder.
2. Right-click (press *Ctrl* and right-click on a Mac) on the `Mapping` folder to open the context menu and select **Create** | **C# Script**. Rename the script `MathG`.
3. Double-click on the new `MathG` script to open it in MonoDevelop or the preferred editor of your choice.
4. Add the namespace by inserting the following line after the using statement:

   ```
   namespace packt.FoodyGO.Mapping {
   ```

5. Remember to close the namespace by adding an end brace `}` to the last line.
6. Change the class definition to match the following:

   ```
   public static class MathG {
   ```

7. Since this is a library, we marked the class static and removed the `MonoBehaviour`.
8. Delete the `Start` and `Update` methods completely.
9. Add new using statement for system, as follows:

   ```
   using system;
   ```

10. At this point, your base library should look like this:

    ```
    using UnityEngine;
    using System.Collections;
    using System;

    namespace packt.FoodyGO.Mapping
    {
        public static class MathG
        {

        }
    }
    ```

If you are new to scripting or scripting in Unity then it is highly recommended that you follow the coding exercise. There is no better way to learn scripting than to just do it. However, if you are an old pro or just prefer to read through the chapters and look at the code later, then that is also fine. Just open the `Chapter_4_Asset` folder from the downloaded source code, and all the completed scripts will be there.

The base new library `MathG` is ready; let's add the haversine distance function, as follows:

1. Create a new method inside the `MathG` class by typing:

   ```
   public static float Distance(MapLocation mp1,
   MapLocation mp2){}
   ```

2. Inside the `Distance` method enter the following first line of code:

   ```
   double R = 6371; //avg radius of earth in km
   ```

3. Next, we enter some more initialization code:

   ```
   //convert to double in order to increase
   //precision and avoid rounding errors
   double lat1 = mp1.Latitude;
   double lat2 = mp2.Latitude;
   double lon1 = mp1.Longitude;
   double lon2 = mp2.Longitude;
   ```

4. After the latitudes or longitudes have been converted from float to double, we calculate the difference and convert the values to radians. Most trigonometric math functions require input as radians and not degrees. Enter the following code:

   ```
   // convert coordinates to radians
   lat1 = deg2rad(lat1);
   lon1 = deg2rad(lon1);
   lat2 = deg2rad(lat2);
   lon2 = deg2rad(lon2);

   // find the differences between the coordinates
   var dlat = (lat2 - lat1);
   var dlon = (lon2 - lon1);
   ```

5. Calculate the distance using the haversine formula by entering the code:

```
// haversine formula
var a = Math.Pow(Math.Sin(dlat / 2), 2) + Math.Cos(lat1) *
Math.Cos(lat2) * Math.Pow(Math.Sin(dlon / 2), 2);
var c = 2 * Math.Atan2(Math.Sqrt(a), Math.Sqrt(1 - a));
var d = c * R;
```

Note that we use the `System.Math` functions, which are double precision to avoid possible rounding errors. Unity also supplies the Mathf library, but that defaults to floats.

6. As you can see, the formula is broken down to a few lines of code for simplicity. There is nothing especially difficult. Finally, we return the value from the function converted back to float and in meters. Enter the last line in the method:

```
//convert back to float and from km to m
return (float)d * 1000;
```

7. Lastly, we need to add a new method to convert the degrees to radians. If you noticed, we used it in the preceding code. Just below the `Distance` method, enter the following method:

```
public static double deg2rad(double deg)
        {
              var rad = deg * Math.PI / 180;
              // radians = degrees * pi/180
              return rad;
        }
```

Now that we can calculate the distance between two mapping coordinates, we need a way to test our formula. Open up the `MonsterService` script in your script editor and perform the following:

8. After the `gpsLocationService` declaration, add the following line:

```
private double lastTimestamp;
```

9. Then, add the following code in the `Update` method:

```
if (gpsLocationService != null &&
    gpsLocationService.IsServiceStarted &&
    gpsLocationService.PlayerTimestamp > lastTimestamp)
{
    lastTimestamp = gpsLocationService.PlayerTimestamp;
    var s = MathG.Distance(gpsLocationService.Longitude,
    gpsLocationService.Latitude,
    gpsLocationService.mapCenter.Longitude,
    gpsLocationService.mapCenter.Latitude);
    print("Player distance from map tile center = " + s);
}
```

10. Most of that code should look familiar by now. The if statement does the same set of tests to make sure that the GPS Location Service is running and has new data points. When it does have new data points, we calculate the distance from the current latitude or longitude and the current map origin. Then, we use print to output the results.
11. One last thing we need to do is add a new using at the top file to include our new `MathG` functions:

```
using packt.FoodyGO.Mapping;
```

12. After you finish editing, make sure that all the scripts are saved and then return to Unity editor. Wait for a few seconds, to let the script updates compile. Make sure that you have the GPS Service set to simulate and then press the Play button to start testing.

    If you jumped here from another part of the book or just forgot how to enable GPS simulation, refer to the section on the *GPS Location Service* in `Chapter 3`, *Making the Avatar*.

13. As the game is running in the editor, open the Unity Console window by selecting **Window | Console**.

14. Drag the Console window tab and attach it to the bottom of the **Inspector** window. Here is the screenshot:

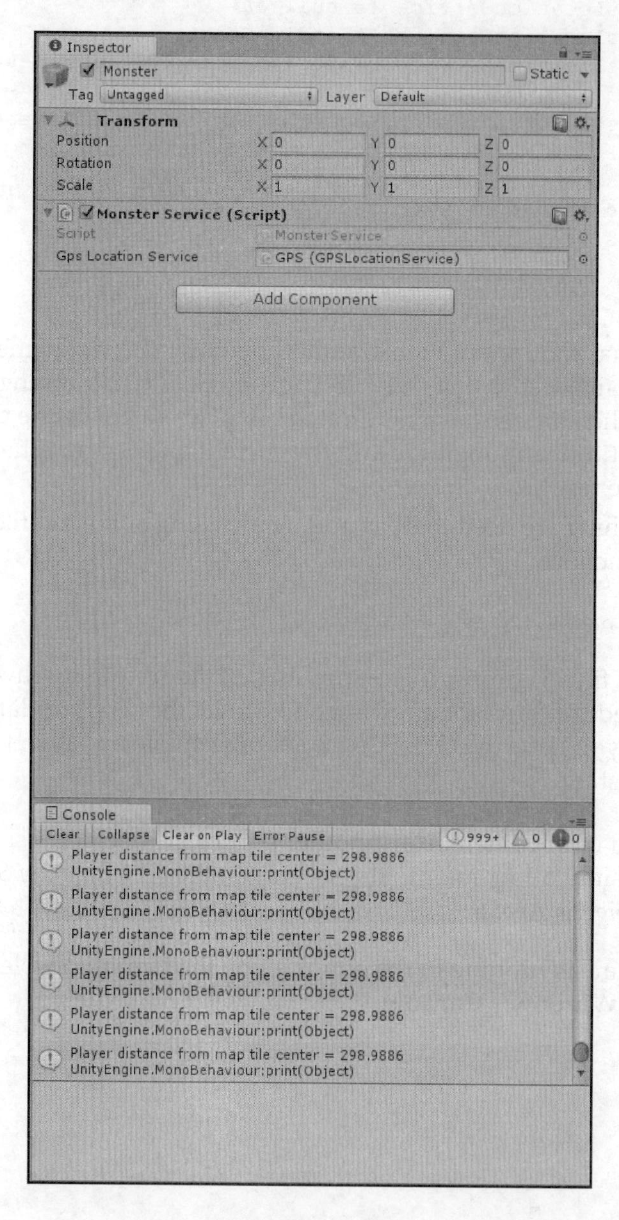

Console window placement

15. You should be seeing the distance the player is from the map tile center in meters. For our current configuration, a single map tile represents an area of roughly 600 x 600 meters. So, our character should never get more than 300 meters away from the center, in any direction before the map redraws itself. If your distances are vastly different, then check the `MathG.Distance` function and make sure that you typed the code correctly.
16. Build and deploy the game to your mobile device. Run the game from your mobile device and attach the CUDLR console. Walk around your house or within Wi-Fi range and see what distances you are getting.

When you walked around your house or the Wi-Fi area, you likely noted that the distances were inaccurate or changed unexpectedly. That inaccuracy is the result of the GPS on your device doing its best to calculate the best possible position using satellite triangulation. As you saw, your device may struggle at times to give an exact location. Since this is important for you to understand as a GPS and mapping developer, we will get into the details of this in the next section.

GPS accuracy

In `Chapter 2`, *Mapping the Player's Location*, when we introduced the concept of GPS tracking, we only briefly mentioned how satellite triangulation worked and what GPS accuracy is. At that time, adding the extra level of detail would have been just information overload and we didn't have a good example to demonstrate GPS accuracy. Now, as you just saw in the last section, GPS accuracy will have an impact on the way the player interacts with the world. As such, let's take a moment to understand how a GPS calculates the location.

A GPS device uses a network of 24-32 satellites orbiting the earth called the **Global Navigation Satellite System (GNSS)** network. These satellites revolve around the earth every 12 hours and transmit their time-encoded location in microwave signals. The GPS device then receives these signals from the satellites it has a clear line of sight to. GPS software on the device uses these readings to calculate distance and use trilateration to find its own location. The more satellites a GPS can see the more accurate the calculation will be.

The following diagram shows how trilateration works:

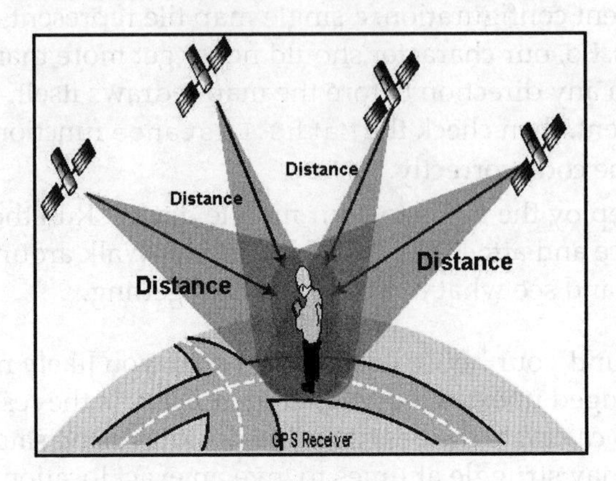

Trilateration of location

Astute readers may have noticed that we switched from using triangulation to using trilateration. That is intentional; triangulation uses angles to determine location. As you can see in the preceding diagram, a GPS actually uses distances. Therefore, the correct and more specific term is trilateration. If you are trying to explain GPS accuracy to your friends though, you still may find the need to use the term triangulation.

At this point, you may be thinking to yourself, aside from the math, if we can see a lot of satellites and can calculate the distances accurately, then our accuracy should be really high. That is true, but there are a number of factors that can skew the calculations. Here is a list of things that can interfere with measuring distances in GPS tracking:

- **Satellite clock offsets**: Satellites use atomic clocks guaranteed to a certain level of precision for the public. Offsets are introduced by the U. S. military in order to intentionally reduce accuracy.
- **Atmospheric conditions**: Weather and cloud cover can impact signals.
- **GPS receiver's clock**: The accuracy of the device's clock will play a major part in accuracy. A mobile phone, for instance, does not have an atomic clock.
- **Signal blockage**: Satellite signals can be blocked by tall buildings, walls, roofs, overpass, tunnels, and so on.
- **Electromagnetic fields**: Power lines and microwaves can have an impact on signal path.
- **Signal bounce**: The worst problem for GPS tracking is signal bounce. Signals can bounce off buildings, metal walls, and more.

The following diagram demonstrates examples of signal problems:

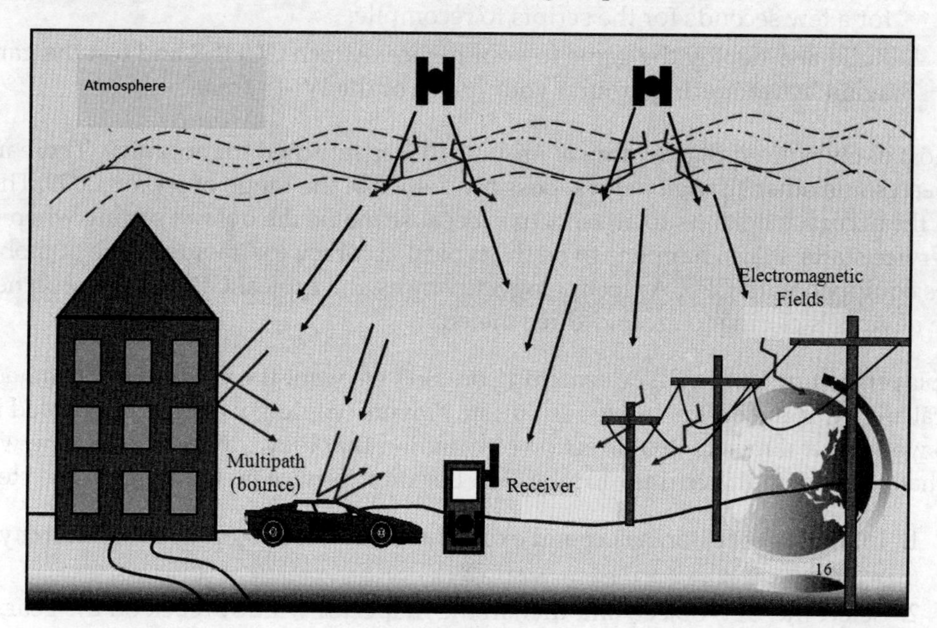

Example of GPS signal problems

Understanding why a device may not be as accurate as it should be can go a long way towards helping you understand problems when you are out testing the game. Try running the game again and wander around your house or the Wi-Fi area with CUDLR connected to your development machine. Check whether you understand why the device may be giving some inaccuracy in the location.

So, now we know that our device is only accurate up to a point. Fortunately, there is an accuracy measure returned from each new location acquisition. This accuracy value will essentially give us the radius of error in meters for the reported calculation. Let's test it out by performing the following:

1. Open the `Assets/FoodyGo/Scripts/Services` folder in the **Project** window and double-click on the `MonsterService` script to open it in the script editor of your choice.
2. Change the print line in the `Update` method to match the following:

```
print("Player distance from map tile center = " + s + "
  - accuracy " + gpsLocationService.Accuracy);
```

3. Save the script when you have done editing and return to the Unity editor. Wait for a few seconds for the scripts to recompile.
4. Build and deploy the game to your device. Attach CUDLR and test the game again by wandering around your house or the Wi-Fi area.

You most likely noticed that a value of around 10 was returned for accuracy. There may have been some other strange values, possibly values in the range of 500 or 1,000. The reason the accuracy defaults to around 10 is because that is the default setting when the GPS service starts. It also happens to be the typical accuracy for a wide range of mobile devices equipped with GPS. As technology advances, the typical GPS accuracy on newer mobile devices is closing to around three meters.

Since players in our game will be searching on foot, we want the GPS updates and accuracy to be calculated at the best possible resolution. Players would get quickly frustrated if they had to walk several meters in a direction only to find that they traveled the wrong way. Let's change the default accuracy for the GPS Location Service by following these steps:

1. In the Unity editor, select and expand the **Services** object in the **Hierarchy** window.
2. Select the **GPS** object, and then in the **Inspector** window click on the gear icon next to the **GPS Location Service (Script)** to open the context menu. Select **Edit Script** from the menu. The following screenshot shows the gear icon and menu:

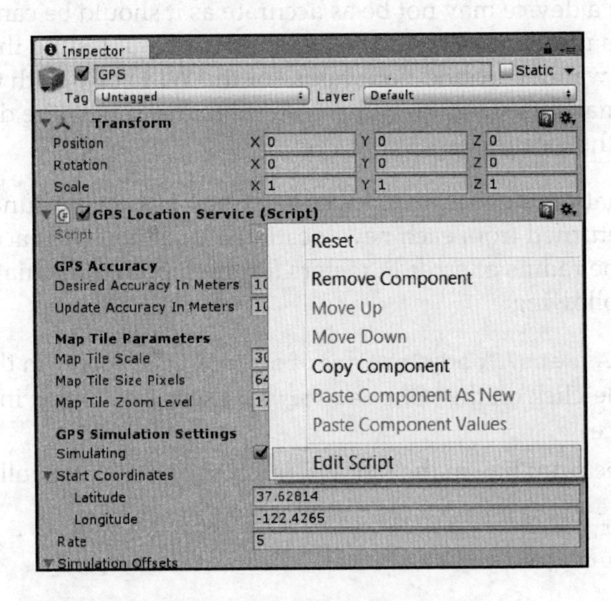

Gear icon context menu to select Edit script

3. You will be directed to the script editor of your choice with the `GPSLocationService` script open.
4. Add the following variable declarations and header attribute to just below the `OnMapRedraw` event declaration:

```
public event OnRedrawEvent OnMapRedraw; //add code after
this line
[Header("GPS Accuracy")]
public float DesiredAccuracyInMeters = 10f;
public float UpdateAccuracyInMeters = 10f;
```

5. Scroll down to the `StartService` method and edit the `Input.Location.Start()` call to this:

```
// Start service before querying location
Input.location.Start(DesiredAccuracyInMeters,
UpdateAccuracyInMeters);
```

6. After you are done with editing, save your script. Return to the Unity editor and wait for the recompilation.
7. Select the GPS object in the **Hierarchy** window. In the **Inspector** window, you should now see the new heading and properties that we just added. Change both of the accuracy settings to 1 from 10.
8. Build and deploy the game to your mobile device.
9. After the game starts running on the mobile device, attach the CUDLR console. Move around your house or Wi-Fi area and check the CUDLR console and note any changes in the results.

Hopefully, your device supports better accuracy than 10 meters, and you may have noticed a flood of new GPS updates come in. If you are seeing those updates, you will also notice that the reported accuracy has likely decreased to a number smaller than 10. For those readers unable to see a change, perhaps you can convince a friend with a newer device to test the game.

Unfortunately, as great as it is having our device now report updates every couple of meters, those more frequent updates come at a price, battery power. The GPS uses a lot of power to constantly receive satellite signals and perform distance and trilateration calculations. By requesting more frequent updates, it is likely the battery on a device will drain quicker. As a game developer, you will need to decide what accuracy works best for your game.

Checking for monsters

Great, now that we understand how to determine distance and how GPS accuracy can alter trilateration of location, it is time to start tracking the monsters around the character. For now, we will use a simple method to randomly place monsters around the player. In a future chapter, we will locate monsters with the help of a web service.

At this point, we have already covered a fair amount of scripting, and we have more to do by the end of the chapter. Also, the scripting changes we need to make are more intricate and prone to mistakes. So, in order to avoid putting you through that turmoil, we will import the next section of changes. For the rest of this chapter, we will switch between manual edits and script imports wherever appropriate. Perform the following instructions to do the first script asset import:

1. From the Unity editor menu, select **Assets | Import Package | Custom Package...**
2. When the **Import package...** dialog opens, navigate to where you placed the book's downloaded source code and open the `Chapter_4_Assets` folder.
3. Select the `Chapter4_import1.unitypackage` file to import and click on the Open button.
4. Wait for the **Import Unity Package** dialog to display. Ensure that all the scripts are selected and click on the **Import** button.
5. Open the FoodyGo folder in the Project window and browse through the new scripts.

Let's open up the new `MonsterService` script in your code editor and take a look at what has changed:

1. Find the `MonsterService` script in the Project window and double-click on it to open it in your code editor.
2. At the top of the file, the first thing you may notice is that some new using statements and a few new fields have been added. Here is an excerpt of the new fields:

```
[Header("Monster Spawn Parameters")]
public float monsterSpawnRate = .75f;
public float latitudeSpawnOffset = .001f;
public float longitudeSpawnOffset = .001f;

[Header("Monster Visibility")]
public float monsterHearDistance = 200f;
public float monsterSeeDistance = 100f;
```

```
public float monsterLifetimeSeconds = 30;
public List<Monster> monsters;
```

3. As you can see, new fields have been added to control monster spawning and at what distances monsters can be seen or heard. Finally, we have a list variable that is holding a new Monster type. We won't spend time looking at the Monster class since it is only a data container at this time.
4. Next, scroll down to the `Update` method and note that the distance test code was removed and replaced with `CheckMonsters()`. `CheckMonsters` is a new method we added to spawn and check the current state of the monsters.
5. Scroll down to the `CheckMonsters` method. The following is the first section of that method:

```
if (Random.value > monsterSpawnRate)
{
    var mlat = gpsLocationService.Latitude +
    Random.Range(-latitudeSpawnOffset, latitudeSpawnOffset);
    var mlon = gpsLocationService.Longitude +
    Random.Range(-longitudeSpawnOffset, longitudeSpawnOffset);
    var monster = new Monster
    {
        location = new MapLocation(mlon, mlat),
        spawnTimestamp = gpsLocationService.PlayerTimestamp
    };
    monsters.Add(monster);
}
```

6. The first line of this method does a check to determine whether a new monster should be spawned. It does this using the Unity Random.value, which returns a random value from 0.0-1.0 and comparing this with the `monsterSpawnRate`. If a monster is spawned, new latitude or longitude coordinates are calculated from the current GPS location and random range +/- the spawn offsets. After that, a new monster data object is created and added to the monsters list.
7. Scroll down a little more, and you will see that the player's current location is converted to a `MapLocation` type. We do this in order to speed up calculations. In game programming, store everything you may need later and avoid creating new objects.
8. In the next line, there is a call to a new Epoch type and storing the result to now. Epoch is a static utility class that returns the current Epoch time in seconds. This is the same time scale Unity uses to return the timestamps from the GPS device.

Epoch or Unix time is a standard for time measurement defined as the number of seconds that have elapsed since 00:00:00 time, which is 1, 1, 1970.

1. Next in the script is a `foreach` loop that checks whether the distance from the monster or player is under the see or hear threshold. If the monster is seen or heard, a print statement outputs the state and distance to the player. The entire remaining section of code is as follows:

```
//store players location for easy access in distance
calculations
var playerLocation = new
MapLocation(gpsLocationService.Longitude,
gpsLocationService.Latitude);
//get the current Epoch time in seconds
var now = Epoch.Now;

foreach (Monster m in monsters)
{
  var d = MathG.Distance(m.location, playerLocation);
  if (MathG.Distance(m.location, playerLocation)
  < monsterSeeDistance)
  {
     m.lastSeenTimestamp = now;
     print("Monster seen, distance " + d + " started at " +
     m.spawnTimestamp);
     continue;
  }

  if (MathG.Distance(m.location, playerLocation) <
  monsterHearDistance)
  {
     m.lastHeardTimestamp = now;
     print("Monster heard, distance " + d + " started at "
     + m.spawnTimestamp);
     continue;
  }
```

2. When you are done reviewing the script, go back to Unity and select the **Monster Service** object in the **Hierarchy** window. Look in the **Inspector** window and review the settings added. Don't change anything just yet.
3. When you are done reviewing the changes, build and deploy the game to your mobile device. Attach CUDLR and do another walk around the Wi-Fi area. As you walk around, check whether new monsters are spawned and the distance.

Great, now we have a way to spawn and track monsters around the player as they move. The obvious next step is to start showing our monsters on the map. However, before we do that, we have to add the code to convert map coordinates to game world coordinates for our Monster service.

Projecting coordinates to 3D world space

If you recall, in the `CharacterGPSCompassController` class the Update method we used already did the conversion from map coordinates to 3D world space. Unfortunately, that code requires a dependency on the GPS Location Service to determine the world map tile scale. So, as much as we would like to create a library function for the conversion, it will just be easier to add it as a helper method to the Monster service.

Fortunately, that helper method was already added as part of the last script asset import. Just go back to your code editor, and assuming you still have the Monster service open from the last section, scroll down to the bottom of the file. You will notice that a private method has been added to do the conversion and is shown as follows:

```
private Vector3 ConvertToWorldSpace(float longitude, float latitude)
{
    //convert GPS lat/long to world x/y
    var x = ((GoogleMapUtils.LonToX(longitude)
        - gpsLocationService.mapWorldCenter.x) *
gpsLocationService.mapScale.x);
    var y = (GoogleMapUtils.LatToY(latitude)
        - gpsLocationService.mapWorldCenter.y) *
gpsLocationService.mapScale.y;
     return new Vector3(-x, 0, y);
}
```

This is the same code we use to convert the player 's coordinates into world space. Essentially, what we do is project the map coordinates to x,y map tile image space and then convert them to world space.

The following figure will hopefully illustrate this concept better:

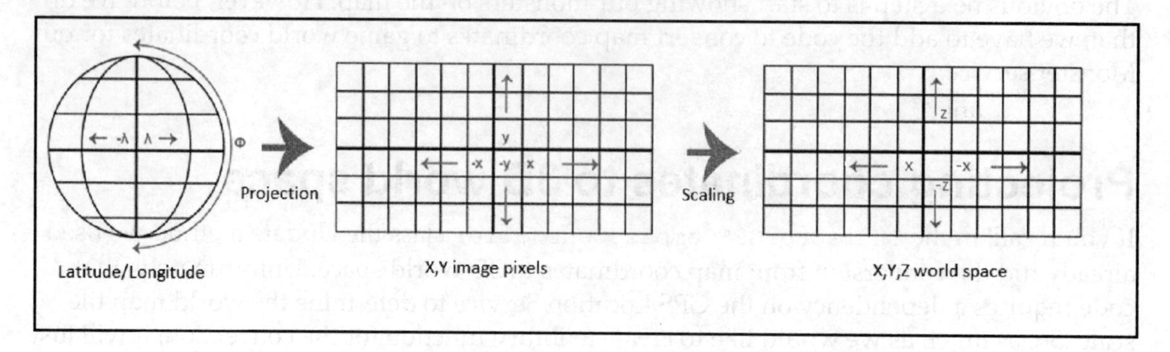

Transformation of latitude or longitude to (x,y,z) world space

Adding monsters to the map

So, now that we have everything lined up, it is time to start actually placing monsters on the map. Well, at least monster objects, to start with. Open up Unity and use the following instructions to add the monster instantiation code to our Monster service:

1. In the search box at the top of the **Project** window, type `monsters`. The assets should automatically filter the items with monsters in their name. Double-click on the **MonsterService** script in the filtered list to open the script in your editor of choice, as shown in the following screenshot:

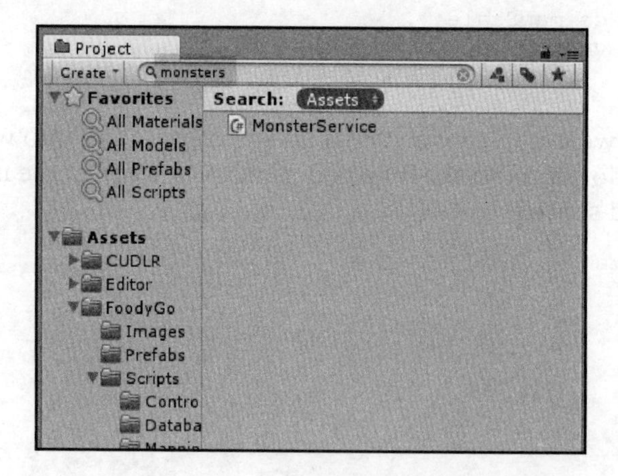

Searching for monsters assets in the Project window

2. Just after the `GPSLocationService` variable declaration, add the following line of code:

```
public GameObject monsterPrefab;
```

3. Scroll down to the bottom of the file, and create a new method `SpawnMonster` with the following code:

```
private void SpawnMonster(Monster monster)
{
    var lon = monster.location.Longitude;
    var lat = monster.location.Latitude;
    var position = ConvertToWorldSpace(lon, lat);
    monster.gameObject = (GameObject)Instantiate(monsterPrefab,
position, Quaternion.identity)
}
```

4. `SpawnMonster` is another helper method we will use to spawn our monster prefab. The `Instantiate` method dynamically creates and returns an object, given a prefab game object and a position or rotation. The returned game object is then added as a reference to the `Monster` data object, which will provide direct access to the game object later.
5. Next, we need to add a call to the `SpawnMonster` inside the `CheckMonsters` method. Locate the following line of code in the `CheckMonsters` method:

```
m.lastSeenTimestamp = now;
```

6. After that line, enter the following lines of code:

```
if (m.gameObject == null) SpawnMonster(m);
```

7. What we are doing here is testing whether the monster already has a spawned object attached. If it doesn't—and it should be visible—we call `SpawnMonster` to instantiate a new monster.
8. Save the script in your code editor and return to Unity. Wait for Unity to do a compile of the updated scripts.
9. Create a new cube game object by selecting **GameObject** | **3D Object** | **Cube** from the menu. Rename the object `monsterCube` in the **Inspector** window.
10. Open the `Assets/FoodyGo/Prefabs` folder in the **Project** window. Then, drag the new `monsterCube` game object to the `Prefabs` folder to create a new prefab.
11. Delete the `monsterCube` game object from the **Hierarchy** window.
12. Select and expand the **Services** object in the **Hierarchy** window, and then select the Monster object.

13. From the `Assets/FoodyGo/Prefabs` folder, drag the **monsterCube** prefab onto the empty **Monster Prefab** slot on the Monster Service component in the **Inspector** window.
14. Press the Play button to run the game in the editor. Ensure that the GPS service is simulating. As the simulation runs, you should see a **monsterCube** object spawn around the player. If you don't see any monsters spawn after a while, lower the monster spawn rate to a number around 0.25 on the **Monster** service. The following is a sample screenshot, notice the how the **monsterCube** clones are added to the **Hierarchy**:

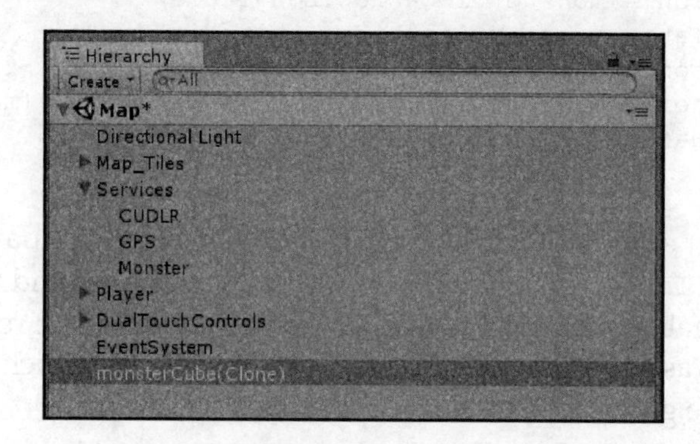

Instantiated monsterCube(Clone) shown in the Hierarchy

Well, obviously, our blocks don't look like very convincing monsters, so let's do something about that. We will use another Reallusion character as the base for our monster. If you recall, Reallusion is the company that creates those great iClone characters we are using for our player character. Perform the following instructions to set up the new monster character:

1. Open the **Asset Store** window by selecting **Window | Asset Store.**
2. After the **Asset Store** page loads, type in the search field `groucho` and press *Enter* or click on search.
3. There will be a paid and free version of the **Groucho** character listed; select the free version in the list.
4. After the asset page loads, click on the **Download** button to download and import the asset. Again, this may take a while; so, grab a beverage or just relax.
5. When the download completes, the **Import Unity Package** dialog will open. Just make sure that everything is selected and click on **Import**, as follows:

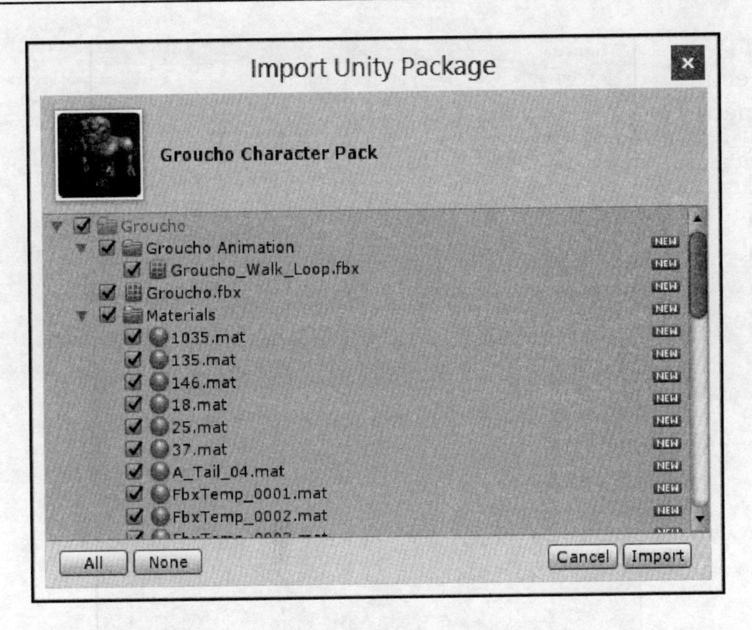

Importing the Groucho character

6. After the character is imported, open the `Assets/Groucho/Prefab` folder in the **Project** window. Then, drag the groucho prefab into the **Hierarchy** window.
7. Select the **Groucho** object in the **Hierarchy** window. In the **Inspector** window, reset the objects transform by selecting the gear icon next to the **Transform** component to open the context menu. Then, select **Reset** from the context menu. The Groucho character should now be overlapping your iClone character.
8. Rename the Groucho object to **Monster** in the **Inspector** window.
9. Click on the target icon beside the **Animation** field of the **Animation** component in the **Inspector** window. Select **Walk_Loop** from the dialog and then close.
10. The imported walk loop animation for the **Groucho** character is not imported to loop by default. We need to fix the animation loop problem by selecting the **Walk_Loop** animation we just set in the **Inspector** window. This will highlight the animation in the **Project** window.
11. Then, select the **Groucho_Walk_Loop** parent object. The animation import properties will then be displayed in the **Inspector** window, as shown in the subsequent screenshot:

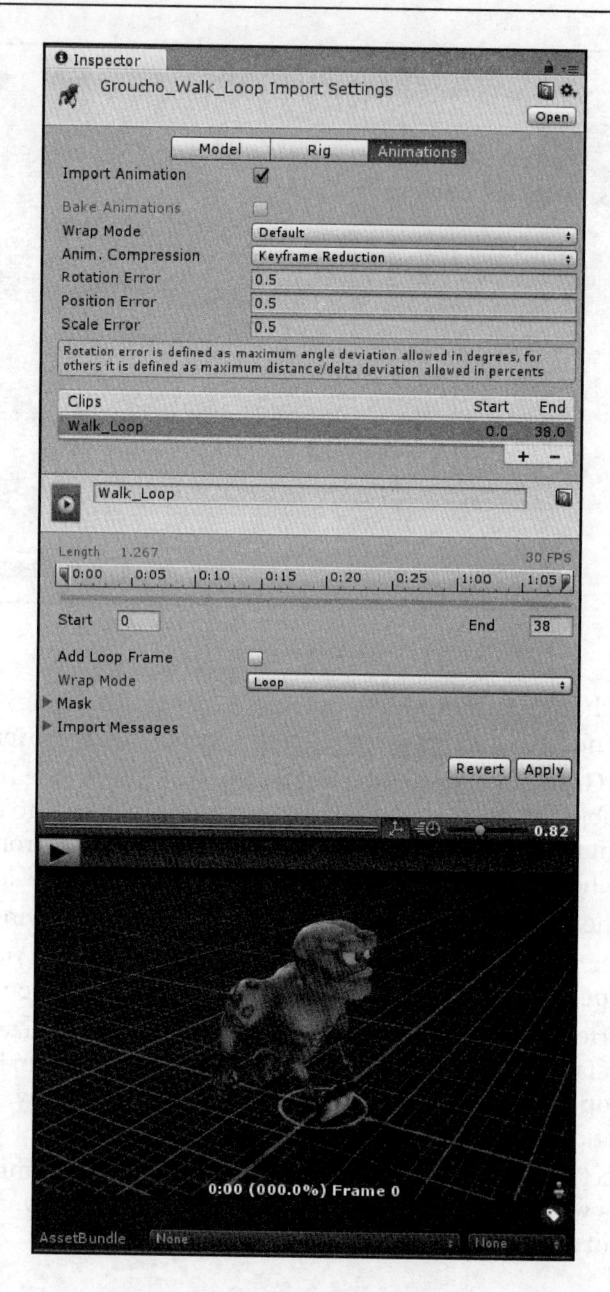

Setting the Wrap Mode of the Walk_Loop animation

12. Change the **Wrap Mode** to **Loop** and then click on the **Apply** button.
13. Select the **Monster** object in the **Hierarchy** window and rescale the monster by changing the scale values in the **Transform** component from 1 to .5 in the x,y,z. The monster needs to be smaller and less intimidating.
14. Drag and drop the **Monster** game object to the `Assets/FoodyGo/Prefabs` folder in the **Project** window to create a new `Monster` prefab.
15. Delete the **Monster** game object from the **Hierarchy** window by selecting and pressing the *Delete* key.
16. Select and expand the **Services** object in the **Hierarchy** window. Then, select the **Monster** service to highlight it in the **Inspector** window. Drag and drop the Monster prefab from the `Assets/FoodyGo/Prefabs` folder in the **Project** window to the **Monster Prefab** field on the **Monster Service** component in the **Inspector** window.
17. Run the game in the editor by pressing Play. Ensure that the GPS simulation is set to simulate. The following sample screenshot shows a high monster spawn rate:

Monsters spawning around the player

The 3D character is designed by Reallusion iClone Character Creator. To create more
customized characters, please visit HERE for more details.

Great, now, we have our monsters spawning on the map. While you were running the game, you likely noted that there are a few new issues. Here is a list of those issue that we need to address:

- When the map recenters itself, the monsters don't recenter
- The monsters remain visible on the map even after a player has moved out of visibility range
- The monsters all face in the same direction
- We still have no way of tracking audible monsters

In order to fix the first three issues, we will do another script asset import and then review the changes. After that, we will fix the final issue by adding a UI element to the scene. Perform the following instructions to import the new scripts and other assets we will need:

1. Import the asset package by selecting **Assets** | **Import Package** | **Custom Package...**
2. After the **Import package...** dialog opens, navigate to the downloaded source code folder `Chapter_4_Assets`, select `Chapter4_import2.unitypackage` and then click on **Open**.
3. When the **Import Unity Package** dialog opens, just confirm that all the assets are selected and then click on **Import**.
4. Confirm whether the new assets are imported by opening the `Assets/FoodyGo` folder in the **Project** window. You should see a some new folders, such as `Images` and `Scripts/UI`.

The first three issues were all fixed with a few additions to `MonsterService` script. Open up the `MonsterService` script in the editor of your choice and review the fixes and respective changes given in the following list:

1. The monsters need to recenter after a map redraw:
    - The first issue was fixed by attaching to the `OnMapRedraw` event of the `GPSLocationService`. If you recall, that event fires when the center map tile redraws itself. Here, you can see the code changes:

```
//event hookup inside Start()
gpsLocationService.OnMapRedraw += GpsLocationService_OnMapRedraw;
```

```
//event method
private void GpsLocationService_OnMapRedraw(GameObject g)
    {
      //map is recentered, recenter all monsters
      foreach(Monster m in monsters)
      {
        if(m.gameObject != null)
        {
          var newPosition =
ConvertToWorldSpace(m.location.Longitude,          m.location.Latitude);
          m.gameObject.transform.position = newPosition;
        }
      }
    }
```

- This method loops through the monsters in the `MonsterService`, checks to see if they have an instantiated game object. If they do, the game object is repositioned on the map.

2. Monsters stay visible after being seen:
    - The next fix is also relatively straightforward with just a couple of additions to the `CheckMonsters` method. The first fix handles when a monster is not seen or heard; we want to ensure that they are not visible. We do this by checking whether a monster's `gameObject` field is not null then we set the `Active` property to false using `SetActive(false)`, which is the same as making the object invisible. Here is the section of that code:

```
//hide monsters that can't be seen
if(m.gameObject != null)
{
  m.gameObject.SetActive(false);
}
```

- Previously, if a monster was seen, we just spawned a new monster if the `gameObject` field was null. Now, we also need to make sure that if the monster does have a `gameObject`, the object is active and visible. We do this almost exactly like we did above, but now we make sure that the game object is active and visible using `SetActive(true)`. The following is the section of code for review:

```
if (m.gameObject == null)
{
            print("Monster seen, distance " + d + " started at " +
m.spawnTimestamp);
```

```
            SpawnMonster(m);
        }
        else
        {
            m.gameObject.SetActive(true); //make sure the monster is
visible
        }
```

3. Monsters all face in the same direction:

    - We will fix this last issue by setting the monsters rotation around the *y* axis of `Up` vector to be random. Here is the section of code, as updated in the `SpawnMonster` method:

```
        private void SpawnMonster(Monster monster)
        {
            var lon = monster.location.Longitude;
            var lat = monster.location.Latitude;
            var position = ConvertToWorldSpace(lon, lat);
            var rotation = Quaternion.AngleAxis(Random.Range(0, 360),
                Vector3.up);
            monster.gameObject = (GameObject)Instantiate(monsterPrefab,
position,      rotation);
        }
```

Tracking the monsters in the UI

For the final issue, we want the player to be able to track monsters that are nearby but not seen. We will do this by providing a visual cue to the player in the form of a footsteps icon or image. One footstep or paw/claw print is very close, two prints not as close, and three prints are just within hearing range. Since we only have one type of monster, at least for now, we will only show the player a single icon representing the closest monster.

Before we get into the code, let's take a look at the new properties that were added to the `MonsterService` in order to support footstep ranges. Expand the **Services** object and then select the **Monster** object. The following is a view of the **Inspector** window for the **Monster** service:

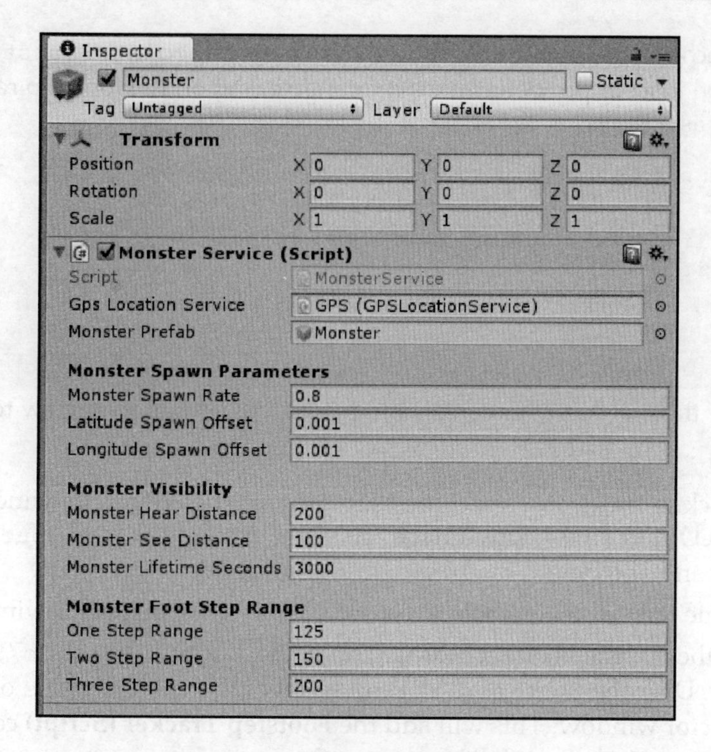

Monster service parameters in the Inspector window

As you can see, in the Inspector window, there is a new section added to the Monster Service component. The new section defines at what ranges the various steps activate, with the value being the maximum range. For example, if the closest monster is at 130 meters distance, the player will see two footsteps because 130 is greater than the 125 set for **One Step Range**, but less than the 150 for **Two Step Range**.

Open up the `MonsterService` script back in your favorite code editor. The following are the script changes where the footstep range is determined and set:

- The first change is in the `CheckMonsters` method inside the `if` statement that checks whether the monster is audible:

```
var footsteps = CalculateFootsetpRange(d);
m.footstepRange = footsteps;
```

- The second change is the addition of the new `CalculateFootstepRange`. This method just simply determines the range based on the footstep range parameters and is as follows:

```
private int CalculateFootstepRange(float distance)
{
    if (distance < oneStepRange) return 1;
    if (distance < twoStepRange) return 2;
      if (distance < threeStepRange) return 3;
      return 4;
  }
```

In order to show the player the footstep's range, we will add an icon view to the UI, as follows:

1. Go back to **Unity** and select the **Map** scene in the **Hierarchy** window. Select **GameObject | UI | Raw Image**. This will expand the `DualTouchControls` object and add a new `RawImage` object as a child.
2. Rename the `RawImage` object to **`Footsteps`** in the **Inspector** window.
3. With the Footsteps object selected, open the `Assets/FoodyGo/Scripts/UI` folder. Drag the `FootstepTracker` script onto the `Footsteps` object in the **Inspector** window. This will add the **Footstep Tracker (Script)** component to the **Inspector** window, as follows:

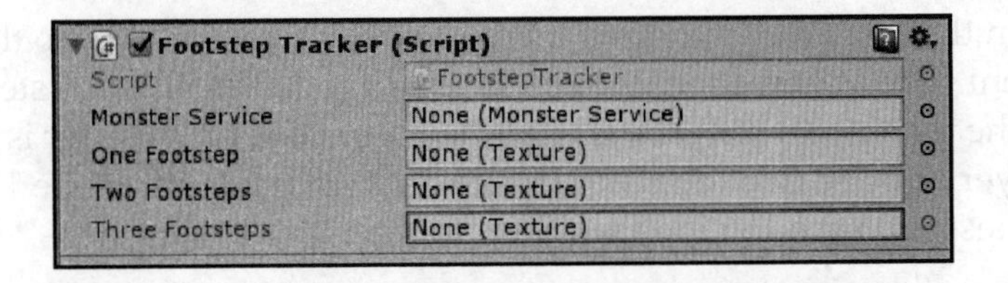

Empty Footstep Tracker (Script) component

4. Expand the **Services** object in the **Hierarchy** window. Drag and drop the **Monster Service** object onto the open **Monster Service** field on the **Footstep Tracker** script component in the **Inspector** window.
5. Click on the bullseye icon beside the **One Footstep** field. The **Select Texture** dialog will open. Scroll down in the dialog and select **paws1** and then close the dialog. This will add the **paws1** texture to the **One Footstep** field.
6. Do the same thing for the **Two Footsteps** and **Three Footsteps** fields, as follows:

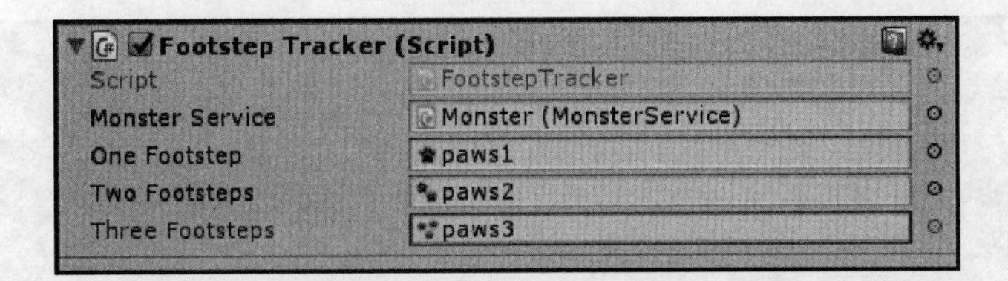

Filled-in Footstep Tracker Script Component

7. Open the **Anchor Presets** menu by clicking on the Rect Transform icon in the **Inspector** window.
8. While the **Anchor Presets** menu is open, hold and press the *Shift* and *Alt* keys and then click on the top left preset, as shown in the next dialog:

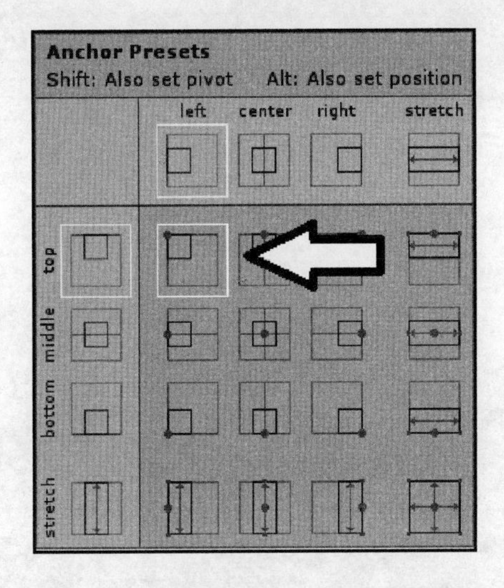

Selecting the Anchor Preset

9. You should now see an empty white square in the top-left corner of the **Game** window. This is where the footsteps icon will appear.
10. Run the game in the **Unity** editor by pressing Play. Ensure that the GPS service is running in simulation mode. As your character moves around now, you should see the footstep's icon turn on with the number of paws representing the distance to the closest monster, as shown in the following screenshot:

One footstep icon showing

After you are done testing the game in the editor, build and deploy it to your mobile device. Then, walk around your house or neighborhood and try to track monsters. Check how close you can get to the monsters. As you are live-testing, be aware of the various distances we set in the Monster Service. Consider whether any of those distance values need to be changed.

Summary

For most of this chapter, we wrote and updated the various core scripts needed to track the monsters on the map. We first started by understanding some fundamental math to calculate spatial distances. Then, we moved on to to understanding GPS accuracy and the factors that influence accuracy. After that, we jumped into scripting the monster service script and other dependent scripts. Next, we added prefab instantiation code to the monster script in order to show a simple prefab on the map. Then, we imported a new monster character into the game and configured a new monster prefab. Through testing, we determined there were a few issues yet to be resolved. These we then resolved through a script update and review. Then, finally we implemented a way to track audible monsters using a simple footstep icon that we added as a new UI element.

In the next chapter, we allow the player to try and catch the monsters inside an alternate reality view. This will be the first chapter where we will explore AR in our game , together with number of other game concepts such as rigid body physics, animation, and particle effects.

5
Catching the Prey in AR

If you recall from our Foody GO storyline, the players needed to hunt down the escaped experimental cooking monsters and catch them. As of the end of last chapter, the player can now track and see monsters around them on the map. What we need now is for the player to be able to interact with and attempt to catch the monsters they can see. In order to keep our game immersion deep, we want the player to catch the monsters in an alternate-reality view. Thus, we also want to incorporate the device's camera to provide the backdrop to our catch activity. Adding the AR component will now classify our game in the real-world adventure or location-based AR genre.

In this chapter, we will be adding a number of new features to our game that will require us to touch on a few new concepts. Unlike the previous chapters, we won't get too heavily into the theory since most of these new concepts are fundamental to game development and Unity. Instead, we will review how these things work within Unity and provide some references for those wanting to learn more about a particular concept. Here is the list of items we will be covering in this chapter:

- Scene management
- Introducing the Game Manager
- Loading a scene
- Updating touch input
- Colliders and rigidbody physics
- Building the AR catch scene
- Using the camera as our scene backdrop
- Adding the catching ball
- Throwing the ball

- Checking for collisions
- Particle effects for feedback
- Catching the monster

Just as in the previous chapters, if you have Unity open from the previous chapter, with the game project loaded, then move on to the next section. Otherwise, open up Unity and load the `FoodyGO` game project or open the `Chapter_4_End` folder from the downloaded source code. Then, ensure that the **Map** scene is loaded.

When you open up one of the saved project files, you will likely also need to load the starting scene. Unity will often create a new default scene rather than trying to guess the scene that should be loaded.

Scene management

Before we jump into adding some new features to the game, we should step back a bit and address how we will transition from scene to scene. Currently, we have two scenes developed for our game: the Splash and Map scenes. Also, in this chapter, we will add two more scenes, Game and Catch. However, we currently have no way of managing scene transitions or game-object lifetime. Ideally, we want some master object and/or script that can do all that for us. This is exactly what we will build, and we will call it Game Manager.

Hold your excitement for a minute. In order to seamlessly load and transition between scenes, we want to do a little housekeeping with our current scenes. Open up Unity and perform the following instructions to clean and reorganize the current game scenes:

1. Ensure that the **Map** scene is loaded in the **Hierarchy** window. Create a new empty game object by selecting the **GameObject** | **Create Empty** menu.
2. Rename the new object **MapScene** and reset the transform to zero.
3. Drag and drop the `Player` object onto the `Map` object in the **Hierarchy** window. This will make the **Player** a child of **Map**. Do this again for **Map_Tiles**, **Services**, **DualTouchControls**, and **Directional Light**, as shown in the following screenshot:

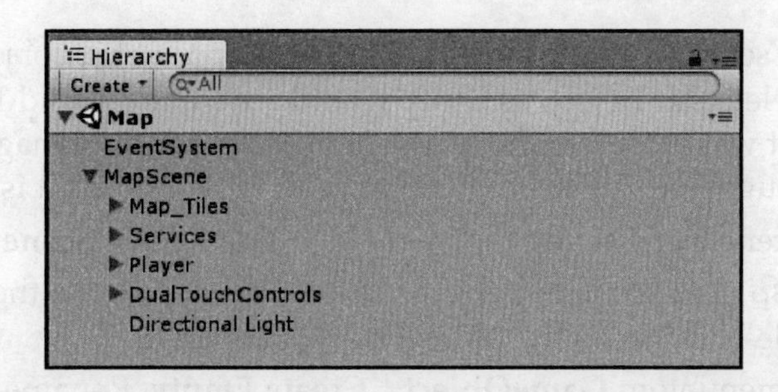

MapScene child objects

4. Save the scene by selecting menu item **File** | **Save Scene**.
5. Save the scene as a new scene called `Game` by selecting menu item **File** | **Save Scene As...**, and in the **Save Scene** dialog, enter `Game` for name and click on the **Save** button.
6. Ensure that you follow the next sequence of steps closely in order to avoid frustration. If you do delete something you shouldn't and accidently save it, then you can always restart from the source code `Chapter_5_Start` folder.
7. In the **Hierarchy** window, delete the `EventSystem` object by selecting the object and pressing the *delete* key. The scene should only contain the `MapScene` and children now.
8. From the menu, select **File** | **Save As...** and name the scene `Map` and then click on the **Save** button. You will be prompted to overwrite the scene; click on **Yes**, as follows:

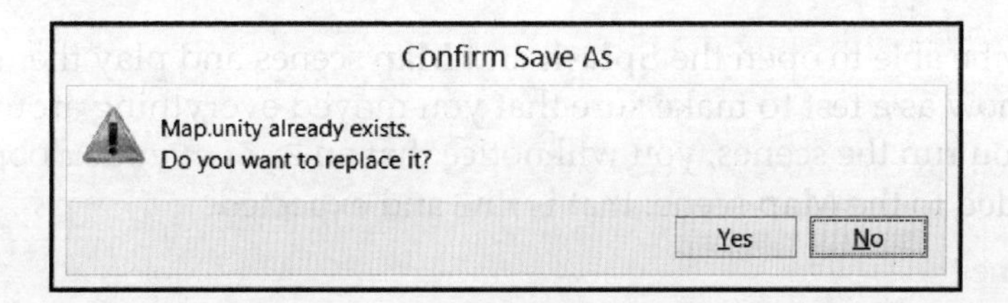

Confirmation prompt

9. Open the new **Game** scene by selecting the `Assets` folder in the **Project** window and then double-clicking on the **Game** scene.
10. Delete the `MapScene` object by selecting it and pressing d*elete*. After you delete the `MapScene` object, you will notice that the **Game** window goes black and now displays a **No cameras rendering** message. Try not to be distracted, panic, or go into repair mode and try to add a new camera to the scene. Everything is OK, so just carry on.

11. The **Game** scene should now just have an `EventSystem` object. We could actually delete this object, as Unity would automatically add it back in if a UI component was added to the scene. In order to better manage our scenes and objects though, we will keep the `EventSystem` object as it is.
12. Save the scene by selecting the menu item **File** | **Save Scene**.
13. Open the `Splash` scene by selecting the `Assets` folder in the **Project** window and then double-clicking on the **Splash** scene to open it.
14. Select the menu item **GameObject** | **Create Empty**. Rename the new object `SplashScene` and reset the transform to zero in the **Inspector** window.
15. Drag and drop the `Main Camera`, `Directional Light`, and `Canvas` to the `SplashScene` object. They should all now be children of the `SplashScene` object.
16. Delete the `EventSystem` object by selecting it and pressing *delete*. The **Splash** scene in the **Hierarchy** window should not look like the following screen excerpt:

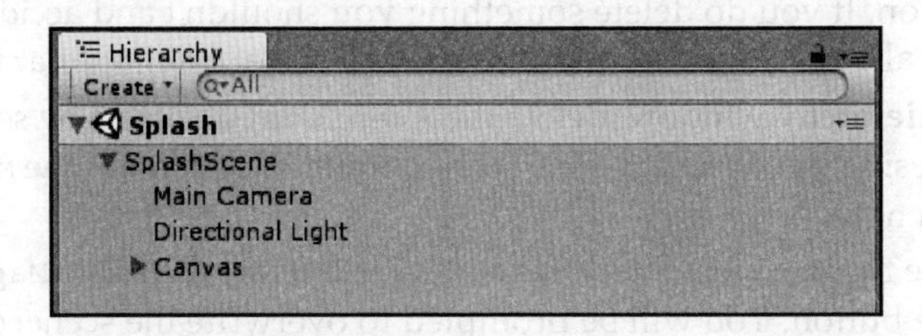

17. Save the scene by selecting menu item **File** | **Save Scene**.

You should now be able to open the **Splash** and **Map** scenes and play them without any issues. Try this now as a test to make sure that you moved everything around and it has all saved fine. As you run the scenes, you will notice that an `EventSystem` object gets dynamically added to the **Map** scene; that is fine and expected.

Introducing the Game Manager

The **Game Manager (GM)** will be our overseer and controller for all the major activities in the game. The GM will manage the scene loading or unloading and transitions, as well as many other higher functions we will get into later. The GM will reside in the Game scene, which will be the first scene loaded. Then, as needed, the GM will manage other activity between scenes, as shown in the following diagram:

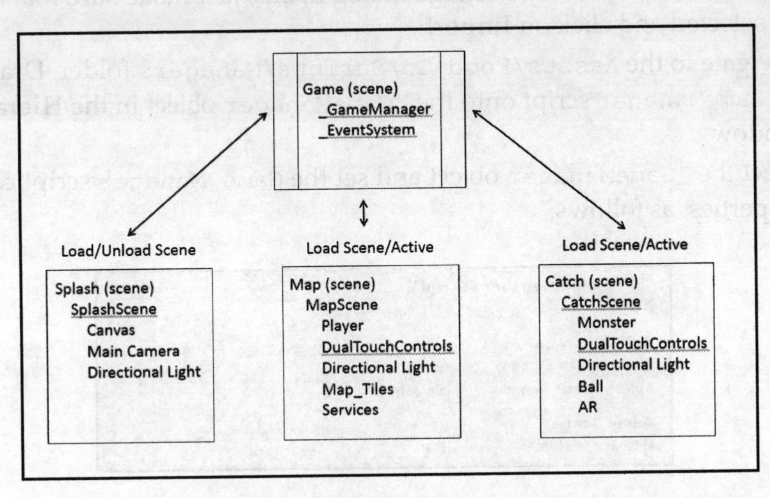

Overview of the scenes and GameManager activity

We will get the `GameManager` game object and script set up and running, and that will explain things further. Unfortunately, this will be a very busy chapter, and we won't have much time to review all the code. It is strongly suggested that you take some time to look over the scripts yourself. Now, follow the next instructions to import and set up the `GameManager` script:

1. Open up the **Game** scene by selecting the `Assets` folder in the **Project** window and double-clicking on the **Game** scene.
2. After the scene loads, create a new empty game object by selecting the menu item **GameObject | Create Empty**.
3. Rename the new object `_GameManager` and reset the transform to zero in the **Inspector** window; note the use of the underscore. We will use an underscore to denote an object that should not be destroyed or deactivated.

4. Select and rename the `EventSystem` object to `_EventSystem` in the **Inspector** window, for the same reason.
5. Select menu item **Assets** | **Import Package** | **Custom Package...**. Then, in the **Import package** dialog, navigate to the downloaded `Chapter_5_Assets` source code folder and select `Chapter5_import1.unitypackage`. Then, click on **Open** to begin the import.
6. When the **Import Unity Package** dialog opens, just make sure that all the assets are selected and click on **Import**.
7. Navigate to the `Assets/FoodyGo/Scripts/Managers` folder. Drag and drop the `GameManager` script onto the `_GameManager` object in the **Hierarchy** window.
8. Select the `_GameManager` object and set the `Game Manager` script component properties, as follows:

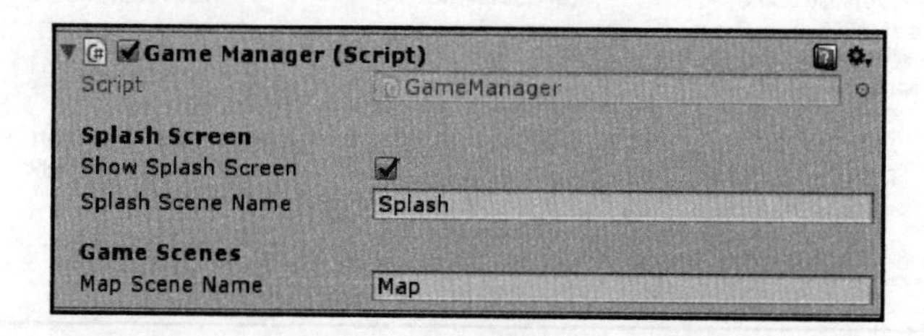

Game Manager settings

9. Select the menu item **File** | **Build Settings...**. We need to add the **Game**, **Splash**, and **Map** scenes into the build settings and set them in the order shown in the following **Build Settings** dialog:

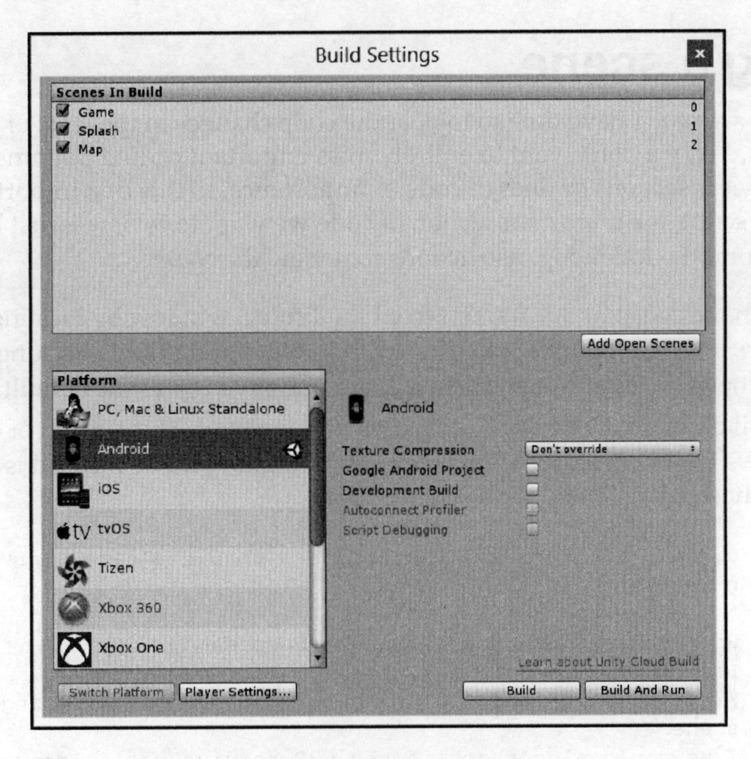

Build Settings dialog, with scenes added and ordered

10. You can add scenes by dragging the scene from the `Assets` folder in the **Project** window and dropping it onto the scene area. The scenes can be reordered in the area by selecting and dragging the scene up or down and dropping it as needed. The first scene in the list will be the scene that gets loaded first. Ensure that your scene configuration matches the dialog image.
11. After you get the scenes added and ordered, run the game in the editor by pressing Play. Notice now that the **Game** scene quickly loads and then the **Splash** scene is loaded, followed a few seconds later by the **Map** scene. You should also notice that the **Map** scene loaded behind the **Splash** scene, in the **Scene** window.
12. Ensure that you also build and deploy the game to your mobile device. Run the game on your device, to be confident that it works as it did from the previous chapter.

Loading a scene

As mentioned, we won't have time to look at the code changes in the level of detail we previously did. Yet, we don't want to entirely miss important coding patterns, which means we will still review sections or lines of code of importance. In this first import of the `GameManager` script, the important section of code we want to review is just how the scene is loaded. Perform the following instructions to review the code:

1. Open the `GameManager` script from the **Project** window by locating the file in the `Assets/FoodyGo/Scripts/Managers` folder and double-clicking on it. This will open the editor of your choice, or `MonoDevelop` as the default.
2. Scroll down to the `DisplaySplashScene` method. This section of code shows the important patterns we want to highlight. The method is shown here for those not able to open a script editor:

```
//display the Splash scene and then load the game start scene
IEnumerator DisplaySplashScene()
{
  SceneManager.LoadSceneAsync(MapSceneName,
  LoadSceneMode.Additive);
  //set a fixed amount of time to wait before unloading splash
  scene
  //we could also check if the GPS service was started and running
  //or any other requirement
  yield return new WaitForSeconds(5);
  SceneManager.UnloadScene(SplashScene);
}
```

3. Inside the coroutine, notice the use of a new `SceneManager` class. The `SceneManager` is a helper class that allows you to dynamically load and unload scenes at runtime. In the first line, the `SceneManager` asynchronously loads a scene in additive rather than the replace mode. Additive scene loading allows you to load multiple scenes together and then unload a scene when it is no longer needed, as shown in the last line of the method.
4. Review the rest of the code for the `GameManager` script on your own. Ensure that you follow how the scenes are loaded and unloaded. You may notice many other things going on in the `GameManager`. Not to worry, we will cover a few of those pieces shortly.

Updating touch input

Now that we can manage our scene transitions through `GameManager`, we now need to set the catalyst that triggers a scene change. For the **Catch** scene, that will be when the player taps on a monster they want to catch, which means we will need to isolate a touch input when it is on a monster. If you recall, our current touch controls work over the entire screen and the input only directs the camera. What we need to do is customize the touch input script to handle that monster touch. Fortunately, for brevity, those script changes were added as part of our last import. Perform the following instructions to configure the new script and review those changes:

1. Open the Unity editor and load the **Map** scene.
2. In the **Hierarchy** window, expand the `MapScene` object and select the `DualTouchControls` object. Rename this object `UI_Input` in the **Inspector** window; this will be a more descriptive name to the function of this object.

Renaming game objects, classes, scripts, or other components to match the function is a good development practice. A good name can be equal to a couple lines of documentation on what the function is, whereas a bad name will cause frustration and upgrade or maintenance nightmares.

3. Expand the `UI_Input` object and select the `TurnAndLookTouchpad`.
4. From the `Assets/FoodyGo/Scripts/TouchInput` folder, drag the `CustomTouchPad` script and drop it on the `TurnAndLookTouchpad` object.
5. This will add the **Custom Touch Pad** script component just below the **Touch Pad** component in the **Inspector** window, as follows:

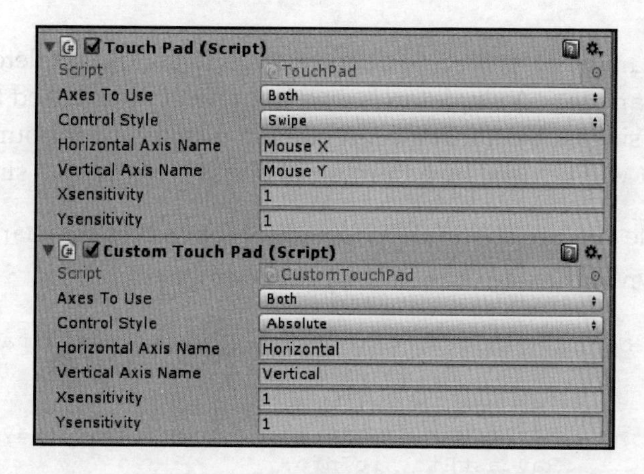

Touch Pad and Custom Touch Pad components in the Inspector window

6. Copy all the settings from the **Touch Pad** component into the **Custom Touch Pad** component. Ensure that both the settings are identical.
7. Remove the **Touch Pad** component by clicking on the gear icon and selecting **Remove Component** from the context menu.
8. The `CustomTouchPad` script is virtually identical to the `TouchPad` script, with only one line of difference. You may be thinking, why didn't we just modify the original script then? The reason we created a new copy of the script and then modified it is to make it our own. That way, if the **Cross Platform Input** asset needs to be upgraded in the future, our custom script changes won't be overwritten.
9. Click on the gear icon on the **Custom Touch Pad** component and select **Edit Script** from the context menu. This will open the script in the editor of your choice.
10. Scroll down to or search for the method `OnPointerDown`; the following is an excerpt of the method and the one line of changed code:

```
public void OnPointerDown(PointerEventData data)
{
if (GameManager.Instance.RegisterHitGameObject(data)) return;
```

11. The `OnPointerDown` method is called when the user first touches the screen and will be the start of a swipe. What we want to do though is not track a swipe if the touch was on an important object. This is what the new line does. The line of code calls the `GameManager.Instance.RegisterHitGameObject` with the touch position. If an important object is touched, true is returned and we return, not allowing the swipe action to start. If instead nothing is hit, the swipe will act as normal.

`GameManager.Instance` denotes a call to get the singleton instance of the `GameManager`. A singleton is a well-known pattern used to maintain a global single object instance. A singleton is perfect for our `GameManager` since it will be used by a number of classes to control a single game state.

12. Now, while you are still in the code editor, open the `GameManager` class again.
13. Scroll down to the `RegisterHitGameObject` method:

```
public bool RegisterHitGameObject(PointerEventData data)
        {
            int mask = BuildLayerMask();
            Ray ray = Camera.main.ScreenPointToRay(data.position);
            RaycastHit hitInfo;
            if (Physics.Raycast(ray, out hitInfo, Mathf.Infinity,
```

```
        mask))
        {
            print("Object hit " +
            hitInfo.collider.gameObject.name);
            var go = hitInfo.collider.gameObject;
            HandleHitGameObject(go);

            return true;
        }
        return false;
    }
```

14. The function of this method is to determine whether a particular touch input has hit an important object in the scene. It does this by essentially casting a ray from the position on the screen into the game world. You can think of a ray as a light pointer, and perhaps the following diagram will help:

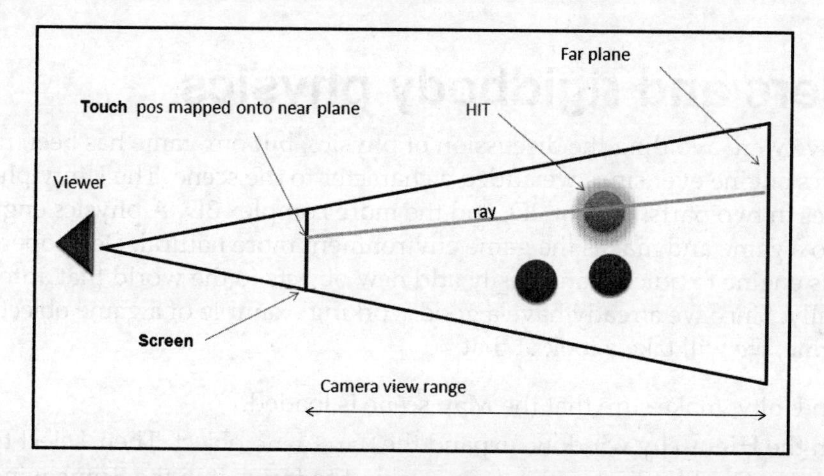

Touch cast as a ray into a scene to determine object's hit

15. The bulk of the work is done within the `Physics.Raycast` method, which uses the ray cast by the touch interaction, a reference to a `RaycastHit` object, the distance the ray should be tested for, and finally, a layer mask to determine whether and how an object was hit. There is a lot going on here, so let's break those parameters down further:
    - `Ray`: This is the ray or the straight line that is cast and used to test for collisions
    - `out RaycastHit`: This returns information about the collision
    - `Distance`: This is the maximum range in which the search should be done

In the code, we use `Mathf.Infinity` for the range of the search. For the number of objects we have in the scene currently, this will work fine. In more complex scenes, a value of infinity could be expensive, as you may only want to test all the objects with in visual distance.

- `Mask`: The mask is used to determine the layers that should be tested for collisions. We will get into more detail about physics collisions and layers in the next section.

16. Go back to Unity and run the game in the editor by pressing Play. As you may notice, nothing new happens. That is because we are missing one vital piece. The `Physics.RayCast` is testing for collisions against objects with colliders. As of yet, our monster object does not have a collider, nor is it in any way configured to use the physics engine, but we will rectify that shortly.

Colliders and rigidbody physics

Until now, we were avoiding the discussion of physics, but our game has been using the Unity physics engine ever since we added a character to the scene. The Unity physics engine comes in two parts: one for 2D, and the more complex 3D. A physics engine is what brings life to a game and makes the game environment more natural. Developers can then leverage this engine to quickly and easily add new objects to the world that automatically react naturally. Since we already have a good working example of a game object using the physics engine, we will take a look at that:

1. In Unity, make sure that the **Map** scene is loaded.
2. In the **Hierarchy** window, expand the `MapScene` object. Then, select the `Player` object. Double-click on the `Player` object to frame it in the **Scene** window.

3. Take a look at the **Scene** window and at the green capsule wrapped around your **iClone** character. In the **Inspector** window, check the **Rigidbody** and **Capsule Collider** components; the following is a screenshot of both windows:

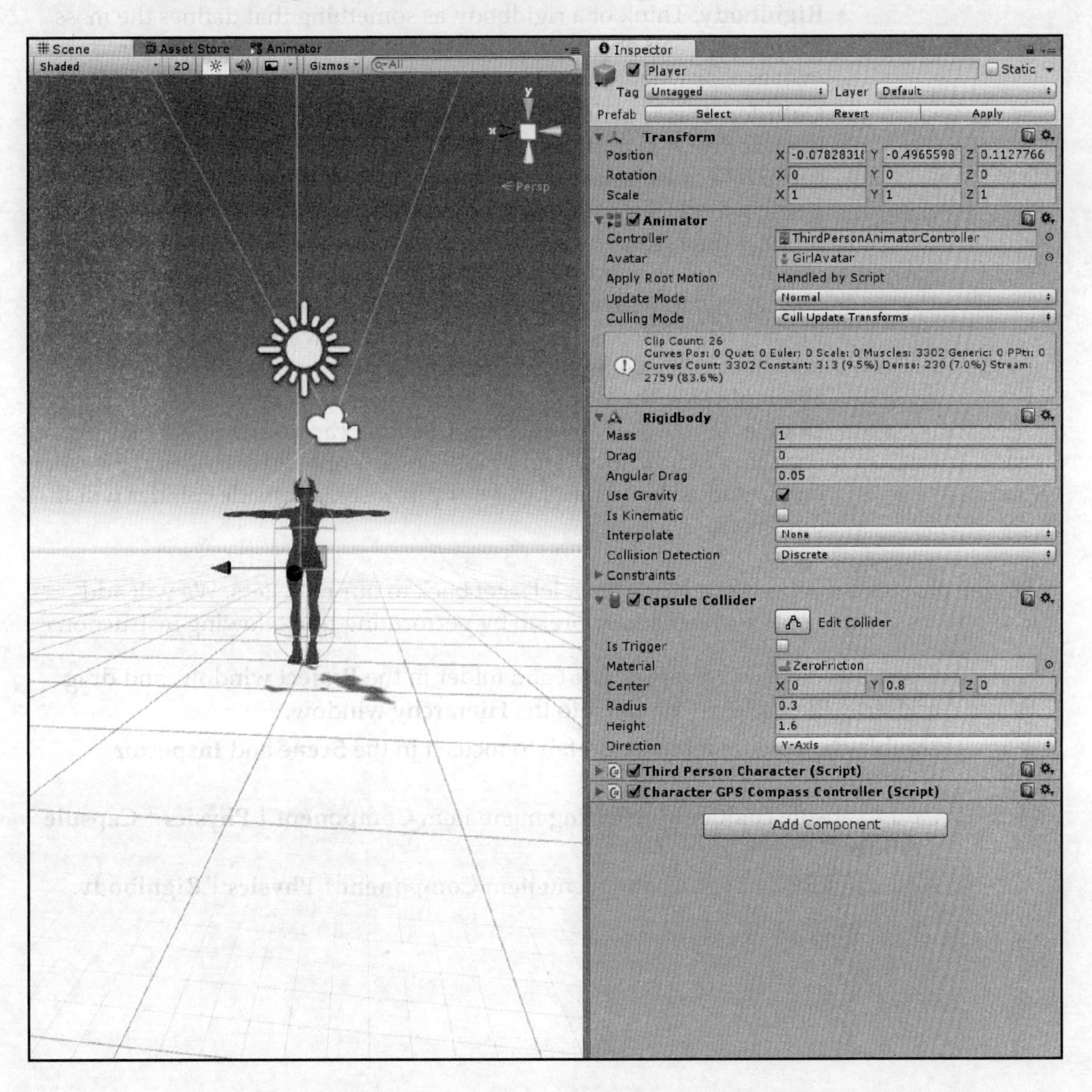

The Scene and the Inspector window showing physics properties of player

4. The rigidbody and collider components are essential to determine the physics of an object; the following is a quick summary of what each of the components does and how they can be interrelated:
    - **Rigidbody**: Think of a rigidbody as something that defines the mass properties of an object, whether the object reacts to gravity, for instance, or what its mass is, how easily it rotates, and so on.
    - **Collider**: This is generally a simplified geometry that defines the object boundaries. A simplified geometry is used, such as a box, sphere, or capsule because collision detection can be performed very quickly. Using a more complex mesh, or even trying to use the actual character mesh, would drag the physics engine to a halt every time a collision was tested. Every frame, the physics engine will test to check whether objects collide with each other. If the objects do collide with each other, the physics engine will then use Newton's laws of motion to determine the effect of the collision. Suffice it to say, if you want to learn further about physics, there are plenty of resources available via your friend Google. In more advanced games, several capsule colliders may be used to wrap the torso and limbs. Thus, it allows for collision detection on individual body parts. For our purposes, a capsule collider will fit our needs.

With the physics definitions out of the way, let's get back to our monsters. We will add those physics components to the monster prefab by performing the following instructions:

1. Open the `Assets/FoodyGo/Prefabs` folder in the **Project** window, and drag and drop the `Monster` prefab into the **Hierarchy** window.
2. Double-click on the **monster** prefab to focus it in the **Scene** and **Inspector** windows.
3. Add a capsule collider by selecting menu item **Component | Physics | Capsule Collider**.
4. Add a rigidbody by selecting menu item **Component | Physics | Rigidbody**.

5. If you look closely in the **Scene** window now, you will see the capsule collider is not wrapping the monster. Try adjusting the **Capsule Collider** component properties by yourself, or use the settings, as shown in the following dialog:

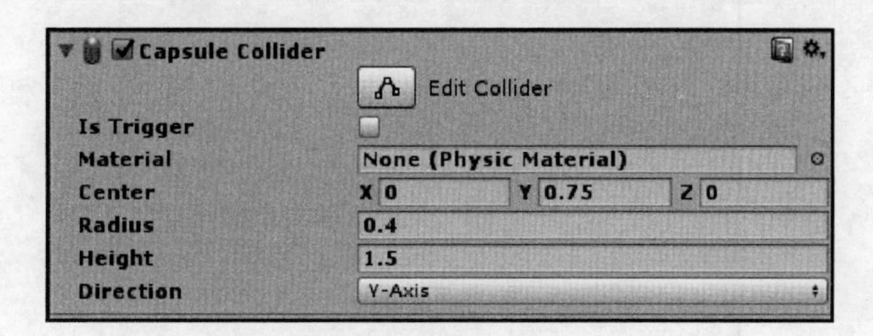

Capsule collider settings for monster

6. With the monster highlighted in the Inspector window, click on the **Apply** button at the top of the window under the **Prefab** actions. This will apply the changes to prefab. Leave the `Monster` object in the **Hierarchy** window.
7. Press Play in the editor to run the game. As you watch the game run, you will hopefully notice that the character now lands on top of the monster and then jumps off. The poor monster unfortunately falls over and then just rolls around. If you don't see this the first time, try running the game a couple of times until you do see this.
8. We won't add functionality to get the monster back on his feet. What we will do, though, is not allow the `Player` to interact with the **Monster** objects. The best real-world analogy to this is our game where we are making the monsters ghosts; they can be seen and heard but cannot be touched.
9. Select the `Monster` object. In the **Inspector** window, select the **Layers** dropdown and then choose **Add Layer...**

10. The **Tags and Layers** panel will open. Add two new layers called `Monster` and `Player` in the list, as shown in the following screenshot:

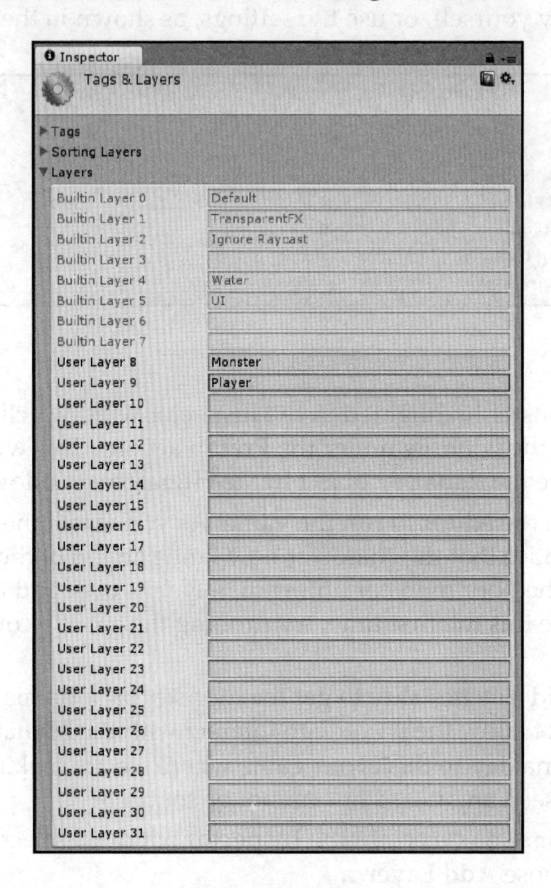

Adding a new Monster layer

11. Select the `Monster` object in the **Hierarchy** window again. The `Monster` will now be on the **Monster** layer shown in the **Layer** dropdown.
12. Select the `Player` object in the **Hierarchy** window. Select the **Layer** dropdown in the **Inspector** window and change the layer to **Player**. You will be prompted to also change the children; click on **Yes, change children**, as follows:

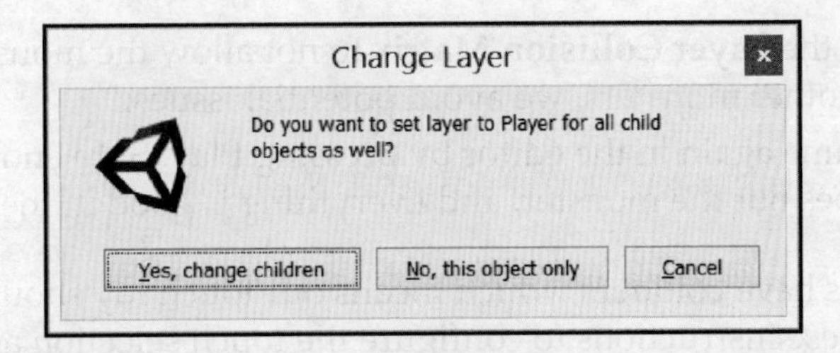

Prompting to change layer on children objects as well

13. Putting the monsters on a new layer will not only allow us to control physics interactions, but also to optimize collision testing. If you recall, the `Physics.Raycast` method took a layer mask parameter. Now that our monsters are on a layer called **Monster,** we can optimize the ray collision test to just the **Monster** layer.
14. Select the menu item **Edit | Project Settings | Physics**. This will open the **PhysicsManager** panel in the **Inspector** window.
15. Uncheck the **Monster-Player** and **Monster-Monster** checkboxes in the **Layer Collison Matrix**:

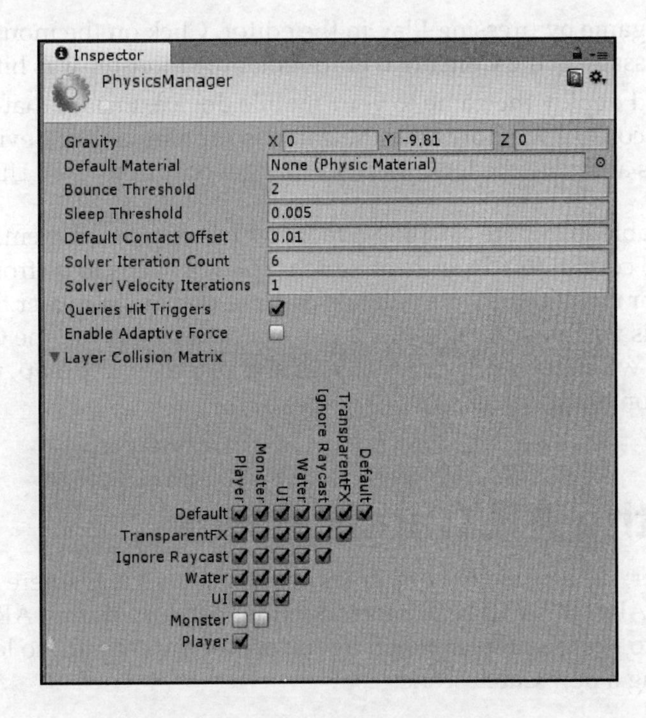

Editing the Layer Collision Matrix in the PhysicsManager

16. By editing the **Layer Collision Matrix** to not allow the monsters to collide with players or other monsters, we avoid potential issues.
17. Run the game again in the editor by pressing Play. Sure enough, our player no longer wipes out the monster, and everything is good in the world again.

So, now our monsters have colliders, which means our touch ray should be able to collide with them. Follow these instructions to configure the touch selection and test out selecting monsters by touch:

1. Select the `Monster` object in the **Hierarchy** window. Click on the **Apply** button on the **Prefab** settings in the **Inspector** window to make sure that all the changes are saved.
2. Delete the **Monster** prefab from the scene by selecting it in the **Hierarchy** window and pressing *delete*.
3. Save the **Map** scene. Then, open the **Game** scene from the `Assets` folder in the **Project** window.
4. Select `_GameManager` in the **Hierarchy** window. In the **Inspector** window, make sure that the **Monster Layer Name** is called `Monster`.
5. Run the game by pressing Play in the editor. Click on the monsters. You should see a message in the **Console** window that the monster was hit.
6. Build and deploy the game to your mobile device. Ensure that you also attach the CUDLR console window after the game is running on the device. Tap on the monsters and note the log messages that get output to the CUDLR console.

The player is now able to initiate catching a monster by tapping on them. Normally, we would also want to control the distance at which a player needs to be from a monster in order to catch it. For now though, we will just assume that any monster that can be seen, may be caught. This will make testing our game easier, especially in the GPS simulation mode. Later, when we start adding other objects and places on the map, we will get into setting an interaction distance.

Building the AR Catch scene

The build up is over; we are finally here: creating the action scene where the players bravely catch the monster. Also, this will be our introduction to incorporating AR into our game. There is lots to do to get this all completed by the end of the chapter, so let's get going. We will start by creating a new **Catch** scene:

1. Create a new scene by selecting menu item **File** | **New Scene**. Then, select **File** | **Save Scene As...** from the menu. In the **Save scene** dialog, enter the name of `Catch` and click on the **Save** button.
2. From the menu, select **GameObject** | **Create Empty**. Rename the new object `CatchScene` and reset its transform to zero in the **Inspector** window.
3. In the **Hierarchy** window, drag and drop the `Main Camera` and `Directional Light` objects on to the `CatchScene` object, to reparent them.
4. From the menu, select **GameObject** | **UI** | **Raw Image**. This will create a `Canvas` object with the `Raw Image` as a child. Select and rename the `RawImage` object to `Camera_Backdrop` in the **Inspector** window.
5. With the `Camera_Backdrop` selected in the **Inspector** window, set the **Anchor Presets** to **stretch-stretch**, by clicking on the Anchors icon to display the menu. Then, hold down the set pivot and set position keys when making the bottom-right corner selection, as follows:

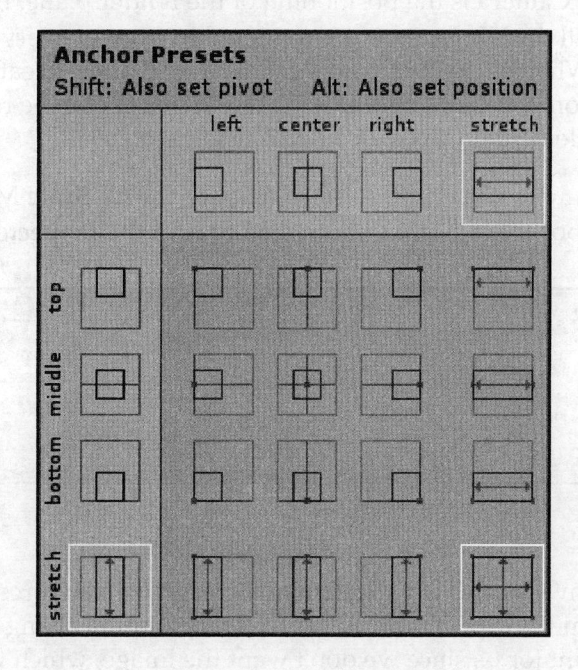

Setting anchor presets on UI element

6. In the **Hierarchy** window, drag and drop the `Canvas` object onto the `CatchScene` object.

7. Select the `Canvas` object. In the **Inspector** window, locate the **Canvas** component. Change the **Render Mode** to **Screen Space – Camera**. Then, drag and drop the **Main Camera** object from the **Hierarchy** window onto the open **Render Camera** slot on the **Canvas** component; refer to the following screenshot for the proper settings:

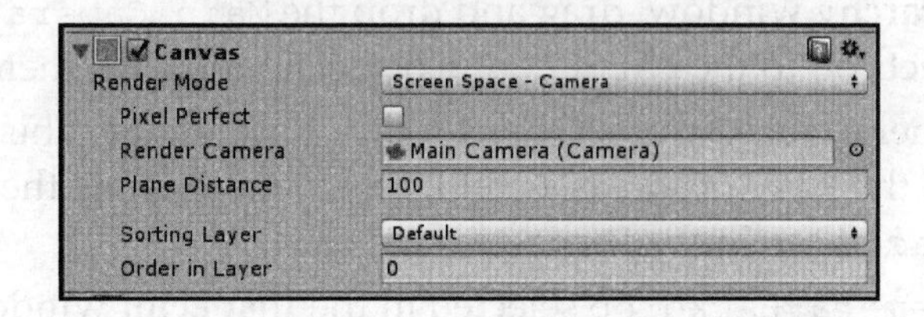

Canvas component settings for Screen Space – Camera

The difference between render modes **Screen Space – Overlay** and **Screen Space – Camera** is the positioning of the render plane. In the Overlay mode, all the UI elements are rendered in front of everything else in the scene. Whereas, in the Camera mode, a UI plane is created a fixed distance away from the camera. This way, world objects can be rendered in front of the UI elements.

8. With the `Canvas` object still selected, change the **UI Scale Mode** in the **Canvas Scaler** component to **Scale with Screen Size** in the **Inspector** window, as follows:

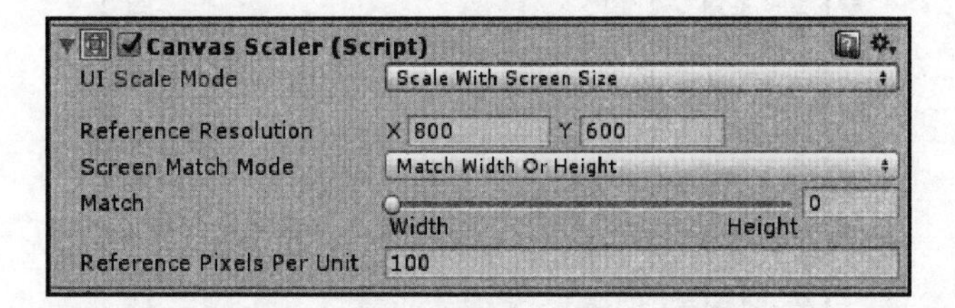

Settings for Canvas Scaler component

The Canvas Scaler with Scale with Screen Size will force the camera to maintain the aspect ratio as the screen resolution changes. This is important for us since we don't want the image, which will be the device's camera, to change aspect ratio or distort.

9. Select and delete the `EventSystem` object. We won't need it here, and even if we do, Unity will create one for us. Finally, save the scene.

That completes the base Catch scene. Time to move on to the world of AR.

Using the camera as our scene backdrop

As you may have already noticed when we created the `Catch` scene, we already put in place a UI element that will act as our scene backdrop. The `Camera_Backdrop` object we set up will display the device's camera as a texture. Follow these instructions to add the script and get the camera view as the scene backdrop:

1. From the menu, select **Assets** | **Import Package** | **Custom Package...**
2. When the **Import package...** dialog opens, navigate to the downloaded source code `Chapter_5_Assets` folder. Select the `Chapter5_import2.unitypackage` and click on **Open** to import the package.

This package is a full import of the FoodyGo assets, and not all of your scripts may need to be updated. However, if you have made your own modifications to some of the scripts, this will overwrite your changes. If you have changes you want to preserve, then either back up the files to a new location or do an asset export of just the files you want to keep.

3. After the **Unity Import Package** dialog loads, review the files you will be importing. There will be a number of new files and some changed files. Just make sure that all the imports are fine and then click on **Import**.
4. Select the `Camera_Backdrop` object in the **Hierarchy** window. In the **Inspector** window, click on the **Add Component** button at the bottom of the window. Select the **Aspect Ratio Fitter** component from the menu.
5. On the **Aspect Ratio Fitter** component, change the **Aspect Mode** to **Height Controls Width.**
6. Click on the **Add Component** button again and select **Camera Texture On Raw Image** component from the menu. You do not need to set any of the fields on this component. The following screenshot shows the proper settings on the new components:

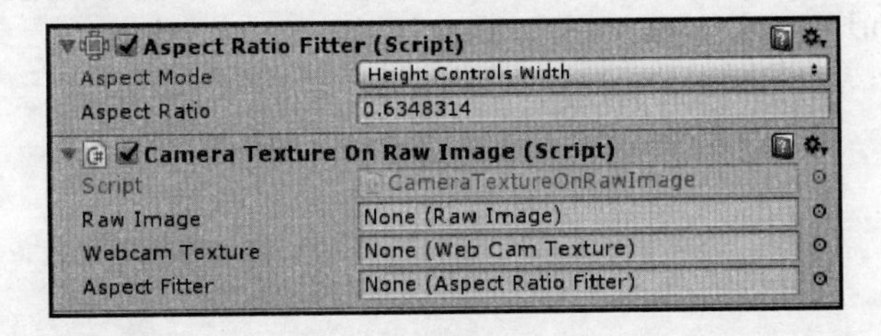

Camera_Backdrop component settings

7. Save the scene by selecting menu item **File** | **Save Scene**.
8. From the menu, select **File** | **Build Settings** to open the **Build Settings** dialog. Click on the **Add Open Scenes** button on the dialog to add the **Catch** scene.
9. Uncheck the **Game**, **Map**, and **Splash** scenes from the build window and check the **Catch** scene, as follows:

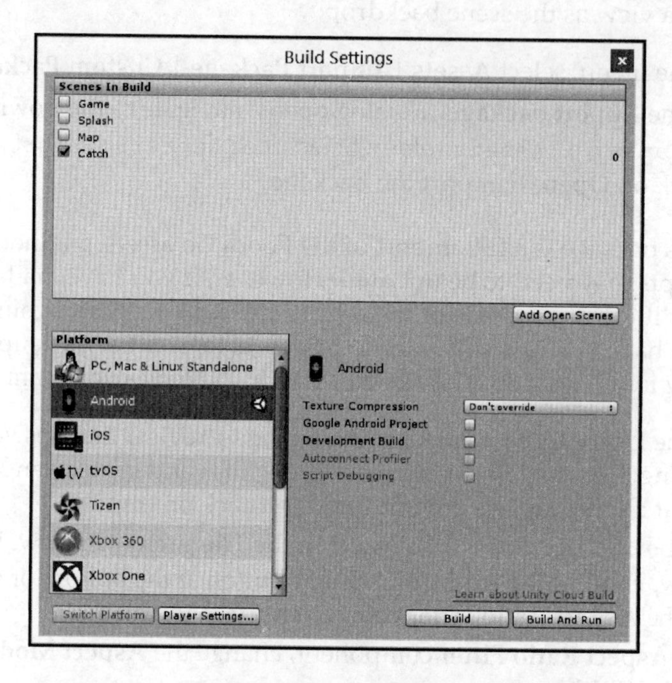

Enabling only the Catch scene in the build

10. Ensure that your mobile device is connected by **USB** to your development machine and then click on the **Build And Run** button. After the game builds and deploys to your mobile device, notice how the background is from the device camera. Rotate the device to change the orientation and notice how the background shrinks; we will fix that in the next section.

As you can see, we are now well on our way to provide the player with an AR experience in the game. With the device's camera providing the scene background, the player will feel that the game objects are in the same area as they are. All of the work for adding the camera backdrop is done in the `CameraTextureOnRawImage` script. Review the script by following this guide:

1. Open up that script in the editor of your choice by clicking on the gear icon beside the **Camera Texture On Raw Image** component in the **Inspector** window. Then, select **Edit Script** from the context menu. Let's concentrate on the code in the `Awake` method:

   ```
   void Awake()
   {
       webcamTexture = new WebCamTexture(Screen.width, Screen.height);
       rawImage = GetComponent<RawImage>();
       aspectFitter = GetComponent<AspectRatioFitter>();

       rawImage.texture = webcamTexture;
       rawImage.material.mainTexture = webcamTexture;
       webcamTexture.Play();
   }
   ```

2. The `Awake` method is called whenever an object is activated, which is typically right after the `Start` method is called. This will be important later when we progress to switching scenes. Otherwise, the code here is fairly straightforward, with main elements getting initialized. The `WebCamTexture` is a Unity wrapper for the webcam or the device camera. After the `webcamTexture` is initialized, it is applied as a texture to `rawImage`. The `rawImage` is the UI background element in the scene. Then, the camera is turned on by calling the `Play` method.
3. We won't dig into the details of the `Update` method in the script here, as it deals with some device orientation artifacts that require manipulation in order to display correctly. The important take away here is that we need to manipulate the camera texture so that it properly orientates to the scene background. Except, if you recall, we still had one issue when the device was rotated to landscape.
4. In order to fix the landscape orientation issue, we will just simply ignore or block the option. Fixing the orientation for all device types and settings is simply beyond the scope of this book. After all, the **Catch** scene will not play well in landscape orientation anyway, so we will force the game to use portrait orientation from now on.

5. Go back to the Unity editor, and from the menu, select **Edit** | **Project Settings** | **Player**. Then, select the **Resolution and Presentation** tab at the top. Inside that tab, look for the **Default Orientation** dropdown and change the default to **Portrait**.

Right now the scene is quite boring, with only the camera running as the background, so let's add our monster to make things interesting:

1. From the menu, select **GameObject** | **3D** | **Plane**. This will create a new plane in the scene. In the **Inspector** window, reset the transform to zero and set the **X** and **Z** scales to 1000. Disable the **Mesh Renderer** component by unchecking the checkbox. This will make the plane invisible.
2. Drag and drop the `Plane` object onto the `CatchScene` object to make it a child.
3. From the `Assets/FoodyGo/Prefabs` folder in the **Project** window, drag the `Monster` prefab into the **Hierarchy** window and drop it onto the `CatchScene` object.
4. In the **Inspector** window, rename the object `CatchMonster` and set the **Transform**, **Rigidbody**, and **Capsule Collider** component properties to match the values, as shown in the following screenshot:

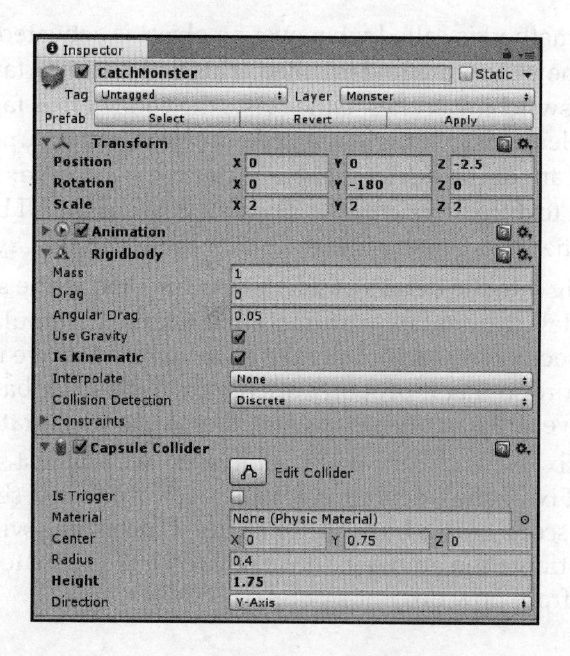

CatchMonster settings in the Inspector window

5. Drag `CatchMonster` into the `Assets/FoodyGo/Prefabs` folder to make a new prefab.
6. Build and deploy the game to your mobile device and watch how the monster now appears inside the game window as a secret window to the world around you.

A few well-known real-world adventure games use a gyroscope-controlled camera in order to add an additional element of realism to the AR experience. A gyroscope-controlled camera will change perspective on the virtual objects as the player orientates their device. We avoided doing this in our game for a few reasons:

- Gyroscope cameras are difficult to code due to the differences in the device's OS and orientation reference. Even on Android, differences may vary by device manufacturer.
- Gyroscope cameras often suffer from some form of drift that needs to be constantly corrected.
- Enabling the gyroscope increases the difficulty in the AR experience. In many other real-world games, players are given the option of disabling AR and they often do so because of the added difficulty. In our case, we want our players to enjoy the AR experience and this is just another reason to not use a gyroscope camera.

In `Chapter 9`, *Finishing the Game* we will talk about other options and solutions to enhance the AR experience we provide to the user. For now though, this simple camera backdrop AR experience will suit our needs.

Adding the catching ball

In our game, the player will use balls of ice to catch the monsters. The player must hit the monster with the ice balls to make them colder and slower so that they will eventually freeze in place. After they are flash frozen, like a frozen dinner, they can then be easily caught.

The ball we will add needs to be made of ice. Since we don't currently have any ice materials to texture a ball with, we will first load a couple of assets. The assets we will load are for particle effects we will use later in the chapter. Conveniently, one of those particle effects also has a nice ice texture we will use for our ball. Perform the instructions here to import the particle effect assets:

1. From the menu, select **Assets** | **Import Package** | **ParticleSystems**. When prompted by the **Import Unity Package** dialog, click on the **Import** button. This will install the standard particle system asset from Unity.

The Unity particle system is known as Shuriken and is the basis for many particle effect assets.

2. Open the **Asset Store** window by selecting menu item **Window** | **Asset Store**.
3. After the window opens and the **Asset Store** page loads, type `elementals` in the search box and then press *enter*.
4. Look for the **Elementals Particle Systems from G.E. TeamDev** asset in the list and select that item. Elementals is an excellent free asset that runs well on most mobile devices.
5. Click on the **Download** button on the asset page to download and import the asset.
6. When you are prompted by **Import Unity Package**, click on the **Import** button to install this asset.

Now that we have the particle system assets imported, we can start adding that catch ball:

1. From the menu, select **GameObject** | **3D** | **Sphere** to add a `Sphere` object to the **Hierarchy** window. Select the new object, and in the **Inspector** window, rename it `CatchBall` and set the transform position and scale to match the following screenshot:

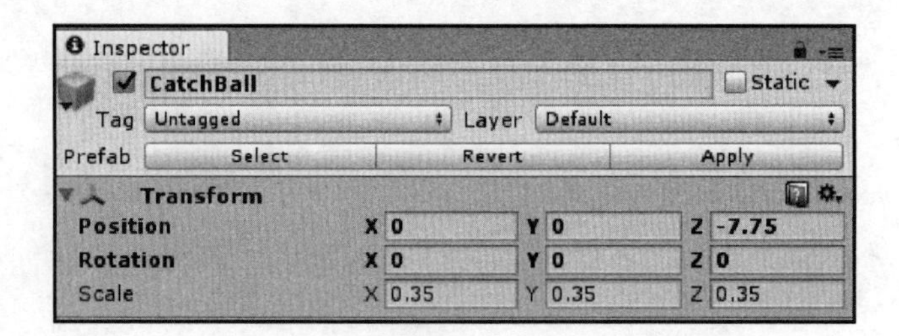

Transform settings for CatchBall

2. Expand the **Materials** list in the **Mesh Renderer** component and select the bullseye icon beside the **Default-Material**. From the **Select Material** dialog, select the **Ice_01** material that points to the `Assets/Elementals/Media/Mobile/Materials/Ice_01.mat`, as follows:

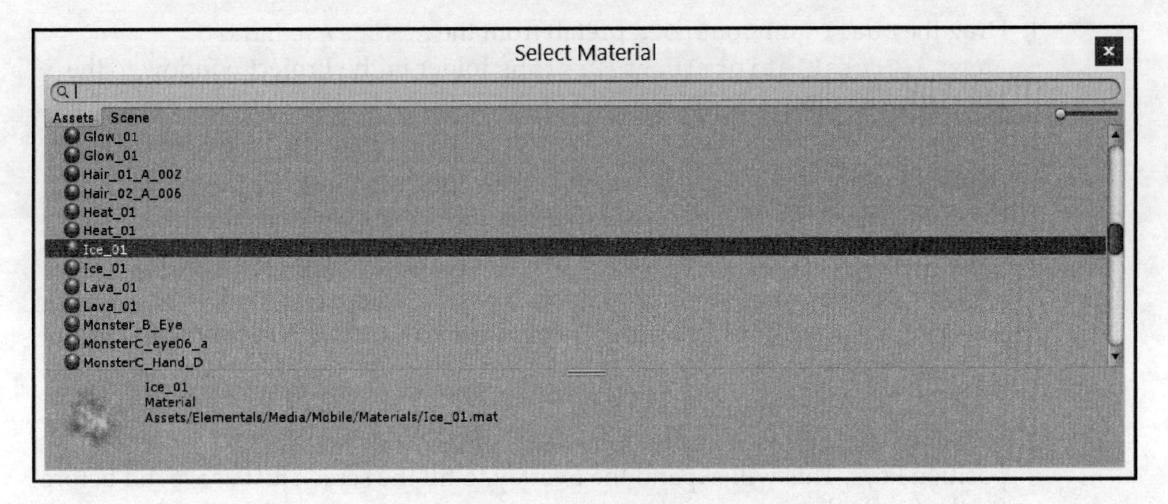

Select the Ice_01 mobile texture variant

3. While you have the **Inspector** window open, click on the **Add Component** button and select **Rigidbody** from the list or enter it in the search field.
4. After the **Rigidbody** component is added to the ball, uncheck the **Use Gravity** field. We will control the gravity on the ball within a script.
5. From the **Hierarchy** window, drag the `CatchBall` and drop it into the `Assets/FoodyGo/Prefabs` folder in the **Project** window. This will create a new `CatchBall` prefab.
6. Again, from the **Hierarchy** window, drag and drop the `CatchBall` onto the `CatchScene` to add it to the scene object.

Throwing the ball

We now have a great-looking ice ball just in front of our walking monster. Time to get that ball in motion using the following steps to add the objects and scripts:

1. Drag the `DualTouchControls` prefab from the `Assets/Standard Assets/CrossPlatformInput/Prefabs` folder in the **Project** window to the **Hierarchy** window.
2. Rename the `DualTouchControls` object to `Catch_UI` in the **Inspector** window.
3. Expand the new `Catch_UI` object and delete the `TurnAndLookTouchpad` and the `Jump` objects. When prompted to break the prefab, click on **continue**.
4. Expand the `MoveTouchpad` object and delete the `Text` object by selecting it and pressing *delete*.
5. Select the `MoveTouchpad` object and rename it `ThrowTouchpad` in the **Inspector** window.
6. Open the **Anchor Presets** in the **Rect Transform** component, and select the stretch-stretch option in the bottom-right corner while holding the pivot and position keys. This will expand the overlay to fill the screen just as we did before for the free look camera.
7. Select the **Image** component **Color** field to open the color dialog. Set the color to `#FFFFFF00` in **Hex**.
8. Click on the **Add Component** button at the bottom of the **Inspector** window. From the drop-down list, select **Throw Touch Pad** or enter it in the search field.
9. Select the gear icon beside the **Touchpad** component, and select **Remove Component** from the context menu.
10. From the **Hierarchy** window, drag the `CatchBall` object and drop it onto the empty **Throw Object** field in **Throw Touch Pad** component, as shown in the following screenshot:

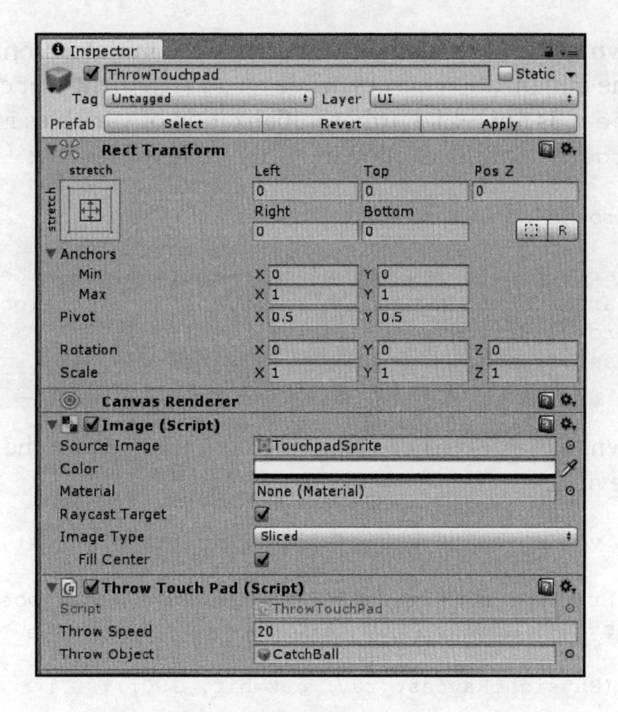

Configuration for the ThrowTouchpad object

11. Drag the `Catch_UI` object onto the `CatchScene` object in order to add it as a child.
12. Press Play in the editor to the run game. You can now click on the ball, drag it in a throwing motion and then release it by unclicking the mouse.
13. Build and deploy the game to your mobile device. Use your finger to throw the ball; see how well you can hit the monster.

If you find it difficult to throw, try to adjust the Throw Speed setting on the Throw Touch Pad component.

All of the work of throwing the ball is being done by the `ThrowTouchPad` script, which uses a heavily modified Touchpad script at its base. Let's review the critical sections of code:

1. Open the `ThrowTouchPad` script in the editor of your choice. Of course, you know how to do this now.

2. Scroll down to the Start method and review the initialization this script does. Most of the initialization occurs inside an `if` statement that checks whether a `throwObject` is not null. Further initialization of variables is down in the `ResetTarget()` call. The following is the code for review:

```
if (throwObject != null)
{
    startPosition = throwObject.transform.position;
    startRotation = throwObject.transform.rotation;
    throwObject.SetActive(false);
    ResetTarget();
  }
```

3. Scroll down to the `OnPointerDown` method; the code for the method is shown here for review:

```
public void OnPointerDown(PointerEventData data)
{
   Ray ray = Camera.main.ScreenPointToRay(data.position);
   RaycastHit hit;

   if (Physics.Raycast(ray, out hit, 100f))
   {
      //check if target object was hit
      if (hit.transform == target.transform)
      {
         //yes, start dragging the object
         m_Dragging = true;
         m_Id = data.pointerId;

         screenPosition =
         Camera.main.WorldToScreenPoint
         (target.transform.position);
         offset = target.transform.position -
         Camera.main.ScreenToWorldPoint(new
         Vector3(data.position.x, data.position.y,
         screenPosition.z));
               }
         }
      }
```

4. This code is very similar to what we used earlier to select the monsters you need to catch. As you can see, that same `Physics.RayCast` method is used, without a layer mask. If the pointer (touch) hits something, it checks whether it is the target, which so happens to be the `CatchBall`. If it has hit the target, then it sets a `Boolean m_Dragging` to true and gets a screen position of the touched object and of the offset of the pointer or touch.
5. Next, if you scroll down a little to the `Update` method, you will see an `if` statement that checks whether the ball is being dragged by checking the `m_Dragging Boolean`. If it is, then it takes a snapshot of the current pointer (touch) position and calls the `OnDragging` method for review:

```
void OnDragging(Vector3 touchPos)
{
    //track mouse position.
    Vector3 currentScreenSpace = new
    Vector3(Input.mousePosition.x, Input.mousePosition.y,
    screenPosition.z);

    //convert screen position to world position with offset
    changes.
    Vector3 currentPosition =
    Camera.main.ScreenToWorldPoint(currentScreenSpace) + offset;

     //It will update target gameobject's current postion.
            target.transform.position = currentPosition;
}
```

6. The `OnDragging` method just moves the target object (ball) around the screen based on the pointer (touch) position.
7. Next, scroll down to the `OnPointerUp` method. The `OnPointerUp` method is called when the mouse button is released or a touch is removed. The code inside the method is quite simple: it again checks whether the `m_Dragging` is true; if it isn't, it just returns. If an object is being dragged, then at this point, the `ThrowObject` method is called, as shown in the following code:

```
void ThrowObject(Vector2 pos)
{
    rb.useGravity = true; //turn on gravity

    float y = (pos.y - lastPos.y) / Screen.height * 100;
    speed = throwSpeed * y;

    float x = (pos.x / Screen.width) - (lastPos.x / Screen.width);
```

```
        x = Mathf.Abs(pos.x - lastPos.x) / Screen.width * 100 * x;

        Vector3 direction = new Vector3(x, 0f, 1f);
        direction =
        Camera.main.transform.TransformDirection(direction);

        rb.AddForce((direction * speed * 2f ) + (Vector3.up *
        speed/2f));

        thrown = true;

        var ca = target.GetComponent<CollisionAction>();
        if(ca != null)
        {
            ca.disarmmed = false;
        }

        Invoke("ResetTarget", 5);
    }
```

8. The `ThrowObject` method is where we calculate the launch position and determine the force with which the object will be thrown. The `x, y` calculations determine how fast the object was moving across the screen before it was released. It determines this by taking the difference between the last-known pointer position and the position of the release. The x value or position of release determines the direction (left or right) of the throw, whereas the y or up movement determines the speed of the throw. These values are then summed into a force vector and applied to the rigidbody by the call `rb.AddForce()`. The `rb` is the target rigidbody, which was set in the `ResetTarget` method during initialization. At the bottom of the method is the call to `GetComponent` for a `CollisionAction` component. We won't worry about this here, but will cover it later. Then, finally, we call `ResetTarget` again using the `Invoke` method, which waits for 5 seconds before being called.

`Rigidbody.AddForce` is one of the most important methods to master when developing any game with physics. You can find more excellent physics resources at `https://unity3d.com/learn/tutorials/topics/physics`.

Checking for collisions

So far, the player can throw the ball at the monster, with little effect. If you managed to hit the monster in your testing, you likely noticed the ball just bounces off, which is certainly not the result we are looking for. What we need now is a way of detecting when the ball hits the monster, or the plane for that matter. Fortunately, the Unity physics engine has a couple of methods to determine when an object collides with another object. Here are the standard options:

- `OnCollisionEnter`: The object has a collider, which makes contact with another game object that also has a collider. The objects will make contact and then push away from each other depending on the force of the collision, and if either or both objects have a rigidbody attached. An object does not need a rigidbody in order to collide, but it does need a collider, as we have seen.
- `OnTriggerEnter`: This occurs when an object has a collider, but the collider is set to be a trigger. A collider set to be a trigger will detect collisions, but then will allow the object to pass right through it. This is useful for things such as doors, portals, or other areas you want to detect when an object enters.

As you have already suspected, we will be using `OnCollisionEnter` to determine when our objects collide. However, instead of writing a single script for each object we want to check for collisions, we will instead implement a collision event system. The problem with writing a collision script for each object is that you will often end up with duplicate code, in different scripts attached to various objects. Each script then often manages its own collisions but often has different rules depending on what object it collides with. Take a look at the following diagram that shows how this may work:

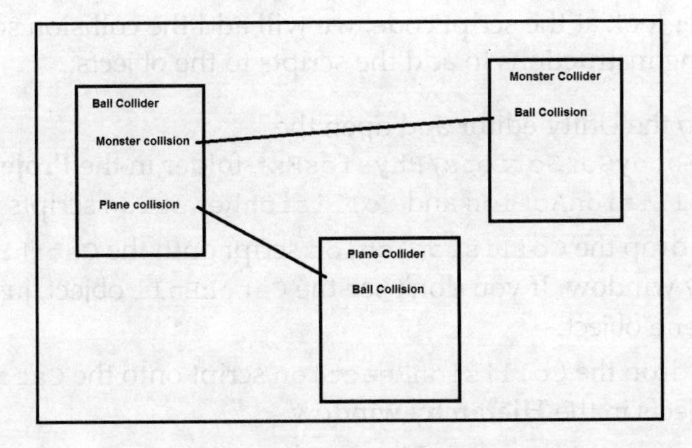

Example of hard-coded collision scripts

As you can see in the diagram, the **Monster** and **Plane** objects both need some of the same code to handle a ball collision. Additionally, the ball object needs to react differently to its hits on the **Monster** or **Plane**. If we add more objects to the scene, we would need to expand that script to account for each object collision. We want a generalized way of acting and reacting to collisions that is also extensible.

Instead of writing three custom scripts in order to manage the collisions in our scene, we will use two, one for the collision (action) and another for the object's reaction. The scripts are aptly named **CollisionAction** and **CollisionReaction**, and a revised diagram showing them attached to the scene objects is as follows:

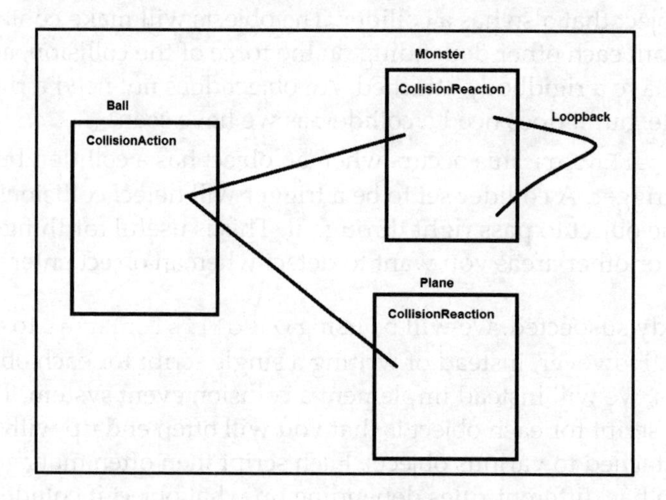

CollisionAction and CollisionReaction scripts used in scene

Before taking a closer look at the script code, we will add the collision scripts to the scene. Perform the following instructions to add the scripts to the objects:

1. Go back to the Unity editor and open the `Assets/FoodyGo/Scripts/PhysicsExt` folder in the **Project** window. You will see the `CollisionAction` and `CollisionReaction` scripts in the folder.
2. Drag and drop the `CollisionAction` script onto the `CatchBall` object in the **Hierarchy** window. If you don't see the `CatchBall` object, just expand the `CatchScene` object.
3. Drag and drop the `CollisionReaction` script onto the `CatchMonster` and `Plane` objects in the **Hierarchy** window.
4. Select the `Plane` object in the **Hierarchy** window.

5. In the **Inspector** window, change the **Collision Reaction** component to match the following screenshot:

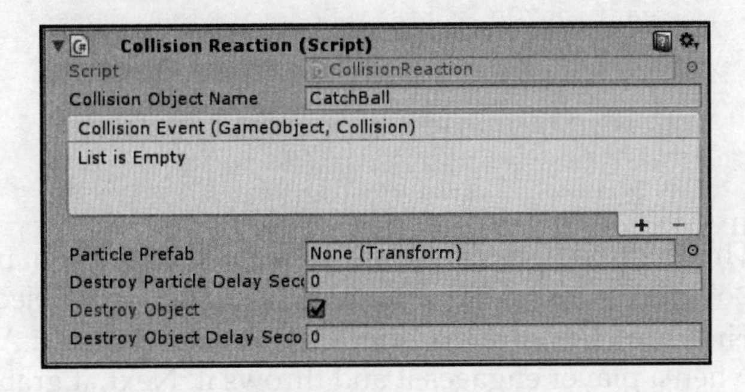

The Collision Reaction component settings

6. Note that we are setting the name of the object (`CatchBall`) we want this component to manage reactions for. By checking **Destroy Object**, we make sure that the ball will now be destroyed when it hits the `Plane`. Don't worry about setting the particle settings just yet; we will get to that shortly.
7. Select the `CatchMonster` object in the **Hierarchy** window and repeat step 4.
8. Run the game in the editor by pressing Play. Test the collision scripts by throwing the ball at the monster and plane. When the ball hits the monster or plane now, it immediately gets destroyed.

As you saw, it was fairly simple to set up those collision scripts. You may expect those scripts to be quite complex, but fortunately they are quite simple. Open up the scripts in the editor of your choice, and we will review them:

- `CollisionAction`: This script is attached to objects that are intended to collide with other objects, such as our ball or bullets and so on. The script detects a collision and then notifies objects with a `CollisionReaction` component that a collision occurred. Let's take a look at the `OnCollisionEnter` method:

```
void OnCollisionEnter(Collision collision)
{
    if (disarmed == false)
    {
       reactions =
       collision.gameObject.GetComponents<CollisionReaction>();
 if(reactions != null && reactions.Length>0)
  {
     foreach (var reaction in reactions)
     {
```

```
                if (gameObject.name.StartsWith
                (reaction.collisionObjectName))
                {
                    reaction.OnCollisionReaction(gameObject, collision);
                }
            }
        }
    }
}
```

This method doesn't actually manage the collision. Instead, it notifies the `CollisionReaction` components of the other object it collided with. First, it makes sure that the object is not disarmed. An object is armed when a player engages it and throws it. Next, it grabs all the `CollisionReaction` components that may be on the collision object. An object may have multiple `CollisionReaction` components attached that react differently for each object. After that, it loops through all the collision reactions and makes sure that object wants to handle collisions by testing the `collisionObjectName`. If it does, then it calls the `OnCollisionReaction` method.

We will use game object names here to filter the object collisions. A better implementation would be to use Tags. It will be up to the reader to be diligent and make that change by themselves.

- `CollisionReaction`: This handles the messy details of what happens when an object collides with it; the code is fairly straightforward:

```
public void OnCollisionReaction(GameObject go,
Collision collision)
{
    ContactPoint contact = collision.contacts[0];
    Quaternion rot = Quaternion.FromToRotation(Vector3.up,
    contact.normal);
    Vector3 pos = contact.point;

    if (particlePrefab != null)
    {
        var particle = (Transform)Instantiate(particlePrefab,
        pos, rot);
        Destroy(particle.gameObject, destroyParticleDelaySeconds);
    }

    if (destroyObject)
```

```
    {
        Destroy(go, destroyObjectDelaySeconds);
    }

    collisionEvent.Invoke(gameObject, collision);
}
```

This is a simple implementation of the CollisionReaction script, but it could be extended or inherited in order to apply other generalized effects, such as decals, damage, and so on.

The first part of the code determines the impact point and direction of the collision. Then, it checks whether a particlePrefab is set. If it is, it instantiates the prefab at the impact point. Then, it calls the Destroy method with a delay defined in the properties. Next, it checks whether the colliding object should be destroyed. If so, it destroys the object after a delay is defined in the settings. Finally, a Unity Event called collisionEvent is invoked, and the collision is passed on to any listeners of that event. This allows for other custom scripts to subscribe to this event and handle the collision additionally, as needed. We will use this event to handle the freezing effects on the Monster later.

- CollisionEvent: At the bottom of the CollisionReaction script, there is another class definition for CollisionEvent : UnityEvent<GameObject, Collision>, which is just empty. This is the definition for a custom Unity Event we use to notify other scripts or components that a collision has occurred. Unity Event is similar to the C# event or delegate pattern we used earlier but is slower in performance. However, Unity Events can easily be wired up between components in the editor rather than hardcoded in a script, which is essential for any script we want to generalize.

Particle effects for feedback

Particle effects in games are like butter to a French chef: essential. They not only provide the pizzazz and special effects we all see in games but are often also subtly used as cues to player activity. For us, we will use particle effects for some pizzazz to enhance scenes and also provide some visual cues. We won't have time to go into the background of particle effects in this chapter, but they will be covered again in `Chapter 8`, *Interacting with an AR World*. Now, let's add some particle effects to our scene:

1. Go back to the Unity editor and open the `Assets/Elementals/Prefabs(Mobile)/Light` folder in the `Project` window.

The Elementals asset is really well done, and you could even try not using the Mobile version by trying the prefabs in the `Assets/Elementals/Prefabs` folder.

2. Select the `Plane` object in the **Hierarchy** window. If you don't see the `Plane` object, remember to expand the `CatchScene` object.
3. Drag the `Holy Blast` prefab from the **Project** window onto the **Particle Prefab** field in the **Collision Reaction** component in the **Inspector** window. Change the **Destroy Particle Delay Seconds** field to 5, as follows:

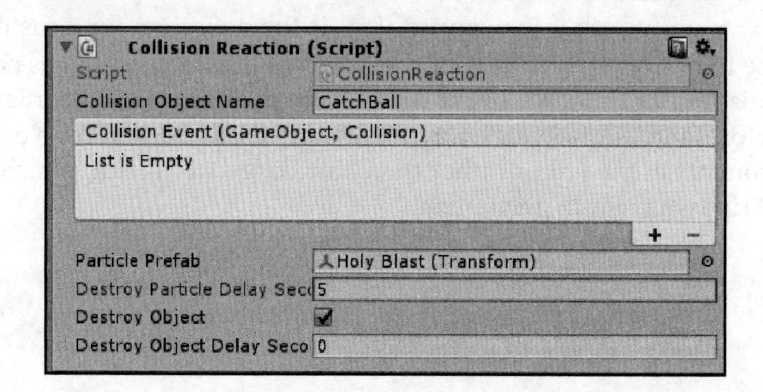

Adding Holy Blast particle prefab to Collision Reaction

4. Select the CatchMonster in the **Hierarchy** window and repeat step 3, exactly the same way.
5. Press Play in the editor window and run the game to test. That certainly looks better now. Build and deploy the game to your mobile device and test there, also.
6. Feel free to try and test other particle prefabs in the **Collision Reaction** components. See what interesting effects are available. You can also experiment with the Unity Standard Assets Particle System located in the Assets/Standard Assets/ParticleSystems/Prefabs folder.

Catching the monster

We have now come to the culmination of our scene: catching the monster. The monster can be hit by the ice balls, and the balls explode on impact, but nothing happens to the monster. If you recall, the player throws the ice balls at the monster in order to freeze them. What we will do then is add a script to slow down our monster with each ice ball hit, and when the monster is hit enough times, it will freeze. Perform the following directions to add and review the script:

1. From the menu, select **GameObject | UI | Canvas**. This will add a new Canvas and EventSystem objects to the scene. Delete the EventSystem and rename the CanvasCaught_UI. Reparent the Caught_UI object to the CatchScene object.

2. Select the `Caught_UI` object in the **Hierarchy** window, and from the menu, select **GameObject** | **UI** | **Text**. Rename the new **Text** object `Frozen`, and set the **Rect Transform** and **Text** component parameters in the **Inspector** window to match the following screenshot:

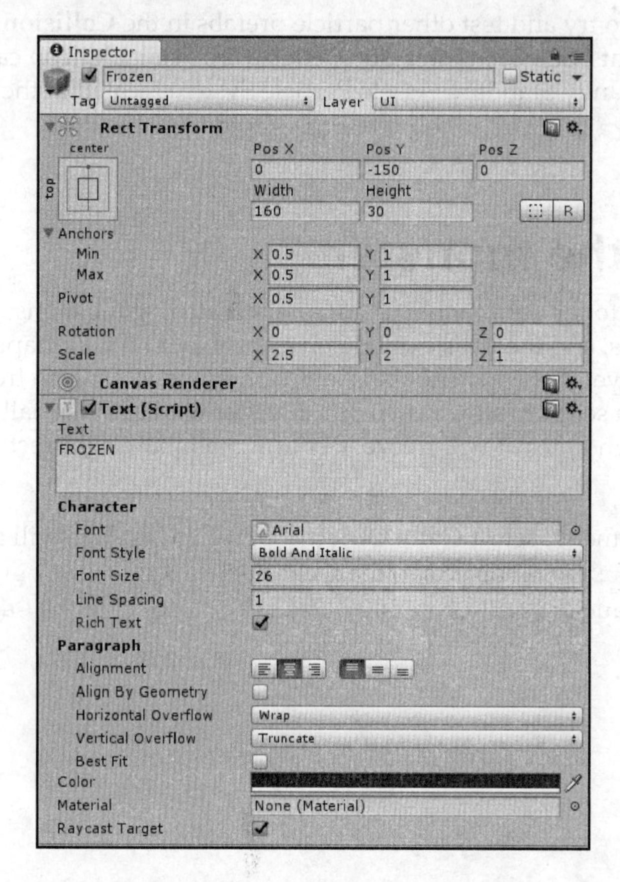

The Frozen component settings

3. Select the `Caught_UI` object in the **Hierarchy** window and disable it by unchecking the checkbox beside the object name in the **Inspector** window. The `FROZEN` text will go invisible in the **Game** window.
4. Select the `CatchScene` object in the **Hierarchy** window. Then, from the `Assets/FoodyGo/Scripts/Controllers` folder in the **Project** window, drag the `CatchSceneController` script and drop it onto the `CatchScene` object.
5. From the `Assets/Elementals/Prefabs(Mobile)/Ice` folder, drag the `Snowstorm` particle prefab onto the **Catch Scene Controller** component's **Frozen Particle Prefab** slot in the **Inspector** window.

6. When you are still in the **Catch Scene Controller** component, expand the **Frozen Enable/Disable List** fields by entering 1 in the **Size** field. Then, from the **Hierarchy** window, drag the `Catch_UI` object to the **Frozen Disable List** and the `Caught_UI` to the **Frozen_Enable_List**:

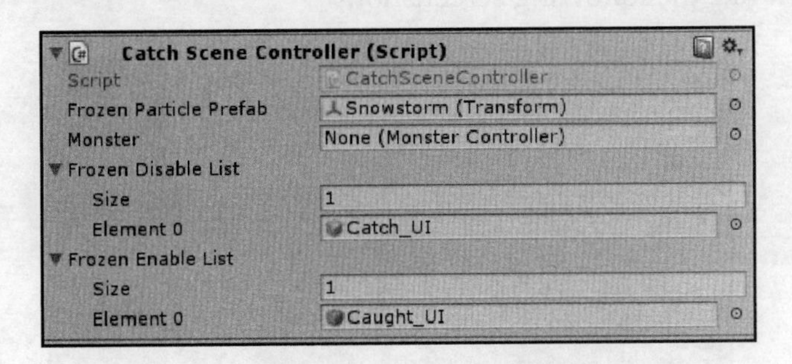

The Catch Scene Controller configuration

7. Select the `CatchMonster` in the **Hierarchy** window. From the `Assets/FoodyGo/Scripts/Controllers` folder, drag the `MonsterController` script onto the `CatchMonster` object in the **Hierarchy** or **Inspector** windows. There is nothing to configure here; adding the script is enough.
8. With the `CatchMonster` still selected in the **Inspector** window, locate the **Collision Reaction** component, likely just above the script you just added. Click on the **+** button under the **Collision Event** field to add a new event listener. From the **Hierarchy** window, drag the `CatchScene` object onto the **None (object)** slot in the new event. Then, click on the dropdown labeled **No Function** to open the context menu and select **Catch Scene Controller | OnMonsterHit**.
9. We just wired up the `Collision Event` | `OnMonsterHit` handler without writing any code. This allows the code we developed to be even more extensible and powerful. Now, if we want to change the game rules and behavior, we only need to modify the `CatchSceneController` script.

10. Press Play and run the game from the editor. Throw balls at the monster, and notice now that as each ball hits the monster, he slows down. Eventually, when enough balls hit the monster he is frozen solid, the FROZEN text will appear and the Snowstorm particles will provide a chilly atmosphere. The **Game** window will look like the following screenshot:

A freshly frozen Monster, served

The 3D character is designed by Reallusion iClone Character Creator. To create more
customized characters, please visit `http://www.reallusion.com/iclone/character-creator/default.html` for more details.

11. As usual, don't forget to test on your mobile device just to ensure that everything runs the same.

Fantastic, we finally have our monster being caught. Before we end this long journey, let's review the last piece of code that brings this all together, the `CatchSceneController`. Open up the script in the editor of your choice and go to the `OnMonsterHit` method, shown here for reference:

```
public void OnMonsterHit(GameObject go, Collision collision)
{
      monster = go.GetComponent<MonsterController>();
      if (monster != null)
      {
         print("Monster hit");
         var animSpeedReduction =
         Mathf.Sqrt(collision.relativeVelocity.magnitude) / 10;
         monster.animationSpeed = Mathf.Clamp01(monster.animationSpeed
         - animSpeedReduction);
         if (monster.animationSpeed == 0)
         {
            print("Monster FROZEN");
            Instantiate(frozenParticlePrefab);

            foreach(var g in frozenDisableList)
            {
                g.SetActive(false);
            }
            foreach(var g in frozenEnableList)
            {
                g.SetActive(true);
            }
         }
      }
}
```

The script is relatively simple and should be easy to follow, but we will identify the highlights:

- The method starts by getting a `MonsterController` from the game object struck in the collision. If the object doesn't have a `MonsterController`, then it isn't a monster and the script just exits.
- After some logging, the method then calculates a damage factor (0-1) and applies that to the monster's animation speed. We won't review the `MonsterController` here, as all it does currently is control the monster's animation speed. The animation speed is clamped to a value between 0 and 1.
- Finally, if the monster's animation speed is zero, the monster is frozen. The `frozenParticlePrefab` is instantiated then the frozen disable/enable lists are looped through and the objects in those list are activated/deactivated.

That wraps up the material for this chapter. If you are concerned that we left certain things hanging, not to worry; in the next chapter, we will continue working on the Catch scene by adding the ability to store the monsters the player catches.

Summary

This was an especially long chapter, but we covered a lot of material and essentially completed a mini-game within our game. First, we talked about scene management, with the loading and transitioning between scenes. A Game Manager was introduced in order to coordinate game activity. Then, we covered touch input, physics, and colliders as part of the player initiating an attempt to catch a monster. This was essential in transitioning us to create a new AR Catch scene. As part of our AR integration, we spent some time understanding how to integrate the device camera into the scene backdrop. We then added the ice ball to our scene and covered how to throw the ball using touch input and physics. After that, we spent some more time discussing colliders and how to script collision reactions. From there, we added the pizzazz to the scene with particle effects, triggered by collision reactions. Finally, we added the catch scene controller script to manage the monster's reaction on being hit. That script also managed the scene objects when the monster was hit enough times and became frozen.

In the next chapter, we will continue where we left off with the Catch scene. We don't want to leave the scene and not be able store the player's catch first. Storing the player's catch and other objects will be essential to our mobile game. Therefore, the next chapter will focus on creating a player database that tracks the player inventory with some new UI elements in order to manage that inventory.

6
Storing the Catch

In the previous chapter, we provided the means for the player to catch a monster by developing the **Catch** scene. However, if you remember, the player could catch a monster, but that was about it. They certainly had nowhere to put the monster after catching it. Well, in this chapter, we will build the player Inventory storage system. This will allow the players to catch and store monsters or other items as needed. Then, of course, we will spend time building out the UI to access the monsters and eventually other items in storage.

For this chapter, we will spend most of the time developing the Inventory system the players will use to store their catch and other items. We will start with the core of our Inventory system, the database. Then, move onto building the UI elements needed by the player in order to access the inventory. Along the way, we will connect the scenes we created earlier and finish up a first working version of our game. Here is a high level overview of what we will cover in this chapter:

- The Inventory system
- Saving the game state
- Setting up services
- Reviewing the code
- Monster CRUD operations
- Updating the Catch scene
- Creating the inventory scene
- Adding menu buttons
- Bringing the game together
- Mobile development woes

Inventory system

If you have ever played any adventure or role-playing game, you are certainly familiar with a player inventory system. The inventory system is an essential element to these games and will be for ours also. Therefore, we will spend a little bit of time reviewing the features we need for our system. The following is a list of features we need for our inventory system:

- **Persistent**: Mobile games are prone to being shut down or interrupted. Therefore, the inventory needs to maintain state between gaming sessions in a database or other methods of storage.

  Saving state should also be robust and quickly executed. For this, we could use a flat file or database. A flat file will generally be simpler to use, but a database is more robust and easily extensible.

You could consider a flat file to be a database. In terms of our discussion, however, we will refer to database as an organized storage mechanism that supports a well-defined data definition and query language.

- **Cross platform**: The underlying database or storage mechanism needs to run on any platform we want to distribute our game on, which at the moment we're targeting Android and iOS.

  The flat file may be an obvious choice for this reason; however as we will see, there are other good cross-platform options as well.

- **Relational**: It's not just in terms of a relational database but in terms of relating objects to one another. For instance, we may want to give our monsters their own items, such as a chef's knife or hat.

  A relational database would work well here but there are other options, such as an object or graph database. Of course, an XML flat file could also represent relationships, but it looks like we are leaning toward a database solution. Our ideal solution would be a relational database that also acts like an object database.

- **Extensible**: Our inventory system will start with a single-item type, the monster. Yet, we will most certainly want to easily support other items later.

  Again, the database wins out here hands-down.

- **Accessible**: Our inventory system will need to be available to multiple parts or scenes within our game.

  Therefore, we likely want our inventory to be a service and/or a singleton type. We could place the inventory system in a service, but it may also make sense to make it a singleton as well.

If you followed along through the features listed in the preceding bullets, it looks like our plan is to use a database. Our preference will be a relational database that is also accessible through objects. Also, we want to use our inventory as a service, like the `Monster` service, but also as a singleton, like the `GameManager` class. Let's take a look at the following diagram that shows how this system should work within the scope of our game:

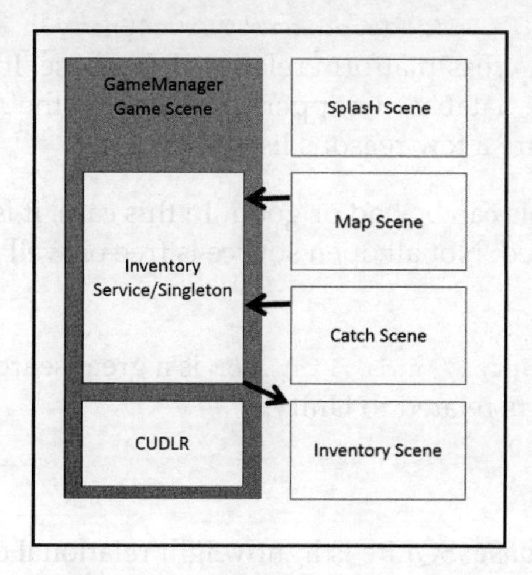

Overview of the Inventory service and communications

As the preceding diagram shows, the new Inventory service will be part of the **Game Scene** and interact with the **Map** and **Catch** scenes. That way, the player will be able to access their inventory from either of those scenes. The inventory UI will all be encapsulated to a new Inventory scene, which will of course be managed by the **Game Manager**.

We almost have everything we need to create our new Inventory service and scene, except for the type and implementation of the database we will be using to drive our inventory system. In the next section, we will get down to picking a database and using it as the core for the new service.

Saving the game state

Our game currently doesn't save any state, but until now we haven't needed it. The player's location was directly accessed from their device's GPS, and the monsters around them were spawned by our makeshift monster service. However, we want our players to hunt down, catch, and collect monsters or other items as part of the game. In order to do this, we need to provide persistent storage in the form of a database. Otherwise, when the player turned off the game, all their collected items would be gone. Games running on a mobile device are particularly susceptible to getting shut down or crashing inadvertently, which means we need a robust storage solution.

If you do a search on the Unity Asset store for database, you will see plenty of free and paid options. Yet, we will use an open source alternative from GitHub called **SQLite4Unity3d** available at `https://github.com/codecoding/SQLite4Unity3d`. This package is an excellent wrapper for SQLite, a great cross-platform relational database. In fact, there are many different versions of SQLite database wrappers available on the Asset store. What puts this software above the others are a few reasons listed here:

- **Open source**: This can be bad or good. In this case, it is good because it is free and still supported. Not all open source is free or well supported, so you have to be careful.

UnityList at `http://unitylist.com` is a great search engine for more open source projects related to Unity.

- **Relational Database**: SQLite is lightweight relational database that also happens to be open source and community driven. The relational database is a good option for us because it supports data relations and provides a well-known data definition language. The language used to query and define data in this database is called a SQL, hence the name. Fortunately, we won't have to have to get into SQL as the SQLite4Unity3d wrapper manages that for us.

SQLite community page is available at `https://sqlite.org/`.

- **Object/Entity Data Model**: An object or entity data model allows a developer to manage data in the database through objects rather than directly writing a secondary language, such as SQL. The SQLite4Unity3d wrapper provides an excellent implementation of a code-first (classes first) object relational mapping or entity-defined data model. A code-first approach will allow us to define our objects first, and then, at runtime, dynamically construct our database to fit our object definitions. Not to worry if this all sounds foreign; we will get into the details of this shortly.

Table first is the opposite of code/classes first approach used to define entities. The database tables are defined first and then the code/classes are derived through a build process. Table first is preferred for those that want a rigorous definition of data.

Now that we have the background out of the way, let's get our hands dirty by loading the database wrapper and other code that we need. Fortunately, the database wrapper and code are all packaged in a single-asset import. Perform the following to import the asset package:

1. Open the Unity editor and continue where we left the project at the end of `Chapter 5`, *Catching the Prey in AR*, with the `Catch` scene loaded. If you have jumped ahead to this chapter, load the project from the downloaded source folder: `Chapter_6_Start`.
2. From the menu, select **Assets** | **Import Package** | **Custom Package...**
3. When the **Import package...** dialog opens, navigate to the downloaded source code folder `Chapter_5_Assets` and select the `Chapter5_import1.unitypackage` file. Then, click on **Open** to import the file.
4. After the **Import Unity Package** loads, just verify what is being imported and then click on the **Import** button. This will import some new and updated scripts as well as some plugins that manage **SQLite** integration.

5. From the **Project** window, select the `Assets/FoodyGo/Plugins/x64` folder. Inside the folder, select the `sqlite3` plugin. Then, in the **Inspector** window, confirm that the plugin is configured to deploy to your device. The following is a screenshot shows an example for **Android**, but it would be the same even for **iOS:**

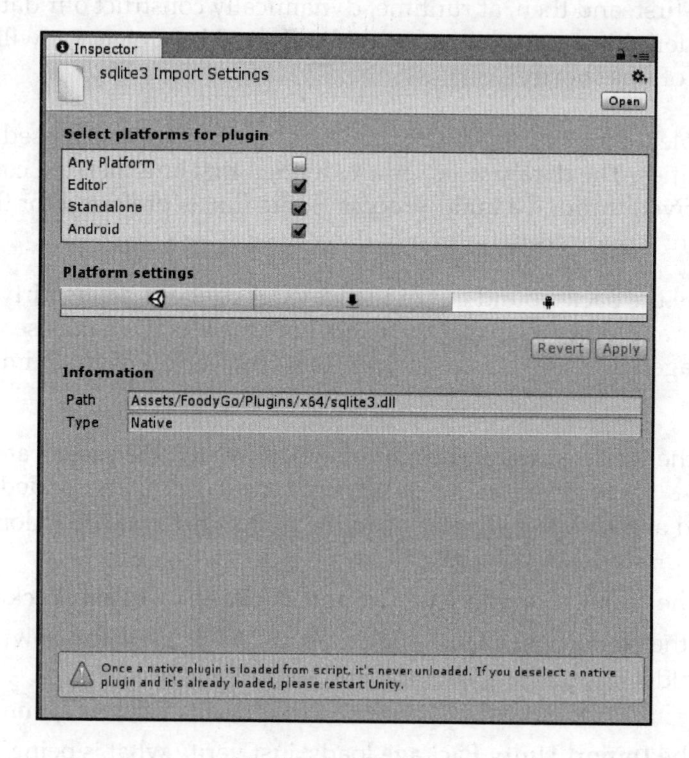

Sqlite3 import settings for Android

6. If you need to make any changes in the panel, click on the **Apply** button at the bottom. This will save and apply the changes.

Importing the package and setting up the plugins were fairly straightforward. We will test whether everything is configured correctly in the next section.

Setting up services

Now that we have the new SQLite wrapper plugins, SQLite script, and other scripts imported, we will set up some services in our `Catch` scene in order to test them:

1. From the menu, select **GameObject** | **Create Empty**. Rename this new object `Services` and reset the transform to zero in the **Inspector** window.
2. Select the new `Services` object in the **Hierarchy** window and right-click (press *control* and click on a Mac) to open the context menu. From the context menu, select **Create Empty**. This will create a new empty child object attached to the `Services` object. Rename this new object **Inventory** in the **Inspector** window.
3. Repeat Step 2, but this time name the new object **CUDLR**.
4. Select the `Inventory` object in the **Hierarchy** window. From the `Assets/FoodyGo/Scripts/Services` folder, drag the `Inventory` script onto the Inventory object.
5. With the `Inventory` object selected, review the **Inventory** component settings in the **Inspector** window, as follows:

Inventory Service component's default configuration

6. The Inventory Service has some critical parameters that are explained here:
   - **Database Name**: This is the name of the database. You should always use a `.db` extension, which is standard for SQLite databases.
   - **Database Version**: This sets the version of the database. The version should always be in the form major.minor.revision and should contain only numeric values. We will discuss how upgrading the database works in detail later.
7. Select the `CUDLR` object in the **Hierarchy** window. From the `Assets/CUDLR/Scripts` folder, drag the Server script onto the `CUDLR` object. If you missed `Chapter 2`, *Mapping the Player's Location*, and the section setting up the CUDLR console, please refer back to that chapter.
8. From the menu, select **Window** | **Console**. Drag and pin the **Console** window to just below the **Inspector** window.

9. Press **Play** in the editor and run the **Catch** scene. You don't have to play the game; just review the output, as shown in the **Console** window:

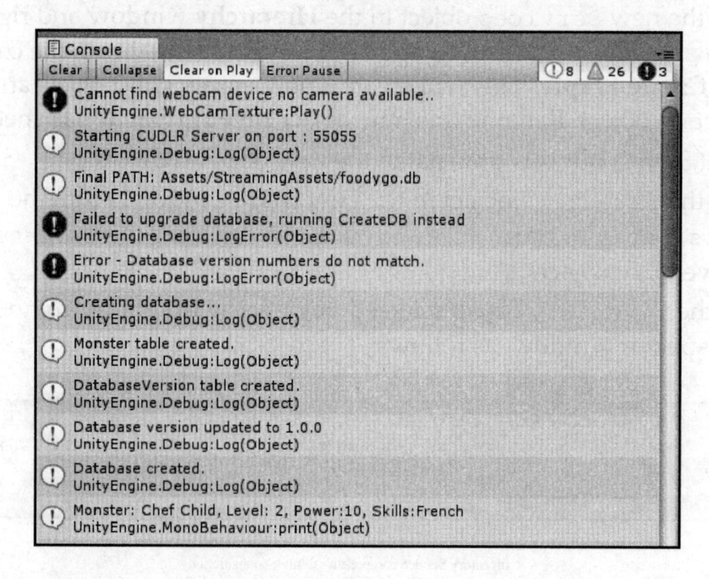

Console output showing a new database being created

10. Your console output should match the output shown in the sample (except for the webcam error). As you can see in the output, the database is actually being created when the game starts.
11. Stop running the game by clicking on the Play button again. Then, run the game again. Note that the **Console** window output is now different. The second time you ran the game, it didn't need to create a database because one was already created.
12. Build and deploy your game to the mobile device. As always, ensure that you have configured the right scenes in your build settings. In this case, only the **Catch** scene should be checked for the build.
13. While the game is running, open your browser and enter the `CUDLR` address you used previously to connect to the console. If you are unsure how to do this, refer back to `Chapter 2`, *Mapping the Player's Location*, on setting up `CUDLR`.

14. The `CUDLR` output should look very similar to the output you see in the **Console** window:

```
Database not in Persistent path
Database written
Final PATH: /storage/emulated/0/Android/data/com.packt.FoodyGO/files/foodygo.db
Creating database...
Monster table created.
DatabaseVersion table created.
Database version updated to 1.0.0
Database created.
Monster: Chef Child, Level: 2, Power:10, Skills:French
```

CUDLR output for creating a new database on Android devices

15. If you don't see something similar on your `CUDLR` output, then check the **Plugin** settings from preceding section or refer to `Chapter 10`, *Troubleshooting*.
16. Close the game on your device. Reopen the game and review the output again on the `CURLR` console. Again, you should not see a new database created, since one is already created for the game.

Reviewing code

As you can see from the previous setup exercise and through testing, the Inventory service we imported already has the database wrapper attached. Let's review some of the script changes we imported and also get into details of the `InventoryService` script:

1. Double-click on the `CatchSceneController` located in the `Assets/FoodyGo/Scripts/Controllers` folder in the **Project** window to open the script in the editor of your choice.

2. The only thing that has thus far changed in the `CatchSceneController` is a new Start method calling our `Inventory` service. Perform the following reviewing method:

```
void Start()
{
    var monster = InventoryService.Instance.CreateMonster();
    print(monster);
}
```

3. Inside the `Start` method, the `InventoryService` is being called as a singleton using the `Instance` property. Then, the `CreateMonster` method is called to generate a new `monster` object. Finally, the `monster` object is printed to the **Console** window with the `print` method.
4. The code in the `Start` method is essentially just temporary test code we will move later. However, appreciate the ease of access the singleton pattern is providing us with.
5. We have one more step before we look at the `InventoryService`. Open the `Monster` script located in the `Assets/FoodyGo/Scripts/Database` folder. If you recall, we previously used the `Monster` class to track our spawn's location within the `MonsterService`. Instead, we decided to simplify the `Monster` class to use just for inventory/database persistence and promote our older class to a new `MonsterSpawnLocation`. The `MonsterService` script was also updated to use the new `MonsterSpawnLocation`.
6. We will take a closer look at the new `Monster` object that we will use to persist to the inventory/database:

```
public class Monster
{
    [PrimaryKey, AutoIncrement]
    public int Id { get; set; }
    public string Name { get; set; }
    public int Level { get; set; }
    public int Power { get; set; }
    public string Skills { get; set; }
    public double CaughtTimestamp { get; set; }
    public override string ToString()
    {
        return string.Format("Monster: {0}, Level: {1}, Power:{2},
Skills:{3}",
                Name, Level, Power, Skills);
    }
}
```

7. The first thing that is obvious is the use of properties to define the `Monster` attributes, which is a divergence from Unity and the more traditional C#. Next, notice that the top `Id` property has a couple of attributes and `PrimaryKey` and `AutoIncrement` attached. If you are familiar with relational databases, you will understand this pattern right away.

 For those less familiar, all records/objects in our database need a unique identifier called a primary key. That identifier, in this case, called `Id`, allows us to quickly locate an object later. The attribute `AutoIncrement` allows us to know that the `Id` property, an integer, will be automatically incremented when a new object is created. This alleviates us from managing the `Id` of the objects ourselves and means the `Id` property will be automatically set by the database.

8. We won't worry too much about the other properties right now, but instead, just take notice of the overridden `ToString` method. Overriding `ToString` allows us to customize the output of the object and is useful for debugging. Instead of having to inspect all the properties and print them to the console, we can instead simplify this to `print(monster)`, as we saw in the `CatchSceneController.Start` method earlier.
9. With the background out of the way, open up the `InventoryService` script located in the `Assets/FoodyGo/Scripts/Services` folder. As you can see, this class has a number of conditional sections in the `Start` method in order to account for various deployment platforms. We won't review that code, but we will take a look at the last few lines of the `Start` method:

```
_connection = new SQLiteConnection(dbPath, SQLiteOpenFlags.ReadWrite |
SQLiteOpenFlags.Create);
Debug.Log("Final PATH: " + dbPath);
if (newDatabase)
 {
 CreateDB();
 }else
 {
 CheckForUpgrade();
 }
```

10. The first line creates a new `SQLiteConnection`, which creates a connection to the SQLite database. The connection is set by passing the database path (`dbPath`) and options. The options provided to the connection are requesting read/write privileges and creating the database, if necessary. Therefore, if no existing database is found at the `dbPath`, then a new empty database will be created. The next line just writes the database path to the Console.

`Debug.Log` is equivalent to the `print` method. We previously used `print` for simplicity and will continue to do so wherever appropriate.

11. After the connection is open, we check whether a new database was created by checking the `newDatabase` Boolean variable. The `newDatabase` variable was previously set above the section of code by determining whether an existing database was already present. If `newDatabase` is true, then we call `CreateDB`, otherwise, we call `CheckForUpgrade`.
12. The `CreateDB` method does not create the physical database file on the device. That is instead done in the connection code we looked at earlier. The `CreateDB` method instantiates the object tables or schema in the database as follows:

```
private void CreateDB()
{
   Debug.Log("Creating database...");
   var minfo = _connection.GetTableInfo("Monster");
   if(minfo.Count>0) _connection.DropTable<Monster>();
   _connection.CreateTable<Monster>();
   Debug.Log("Monster table created.");
   var vinfo = _connection.GetTableInfo("DatabaseVersion");
   if(vinfo.Count>0) _connection.DropTable<DatabaseVersion>();
   _connection.CreateTable<DatabaseVersion>();
   Debug.Log("DatabaseVersion table created.");

   _connection.Insert(new DatabaseVersion
   {
         Version = DatabaseVersion
     });
     Debug.Log("Database version updated to " + DatabaseVersion);
     Debug.Log("Database created.");
   }
```

13. Don't be put off by the number of `Debug.Log` statements in this method; it is best to just think of them as helpful comments. After the install logging, we first determine whether the `Monster` table has already been created using the `GetTableInfo` method on the connection. `GetTableInfo` returns the columns/properties of the table; if no columns or properties have been set, `minfo` will have a count of `0`. If the table is present, however, we will delete or drop it and create a new table using our current `Monster` properties.

    We follow the same pattern for the next table `DatabaseVersion`. If `GetTableInfo` returns `vinfo.Count > 0`, then delete the table, otherwise, just continue. You will see, as we add more objects to the `InventoryService`, we will need to create the new tables in the same manner.

    The SQLite4Unity3d wrapper provides us with an **Object Relational Mapping (ORM)** framework that allows us to map objects to relational database tables. This is why we will use the term object and table interchangeably at times. The following diagram shows how this mapping typically works:

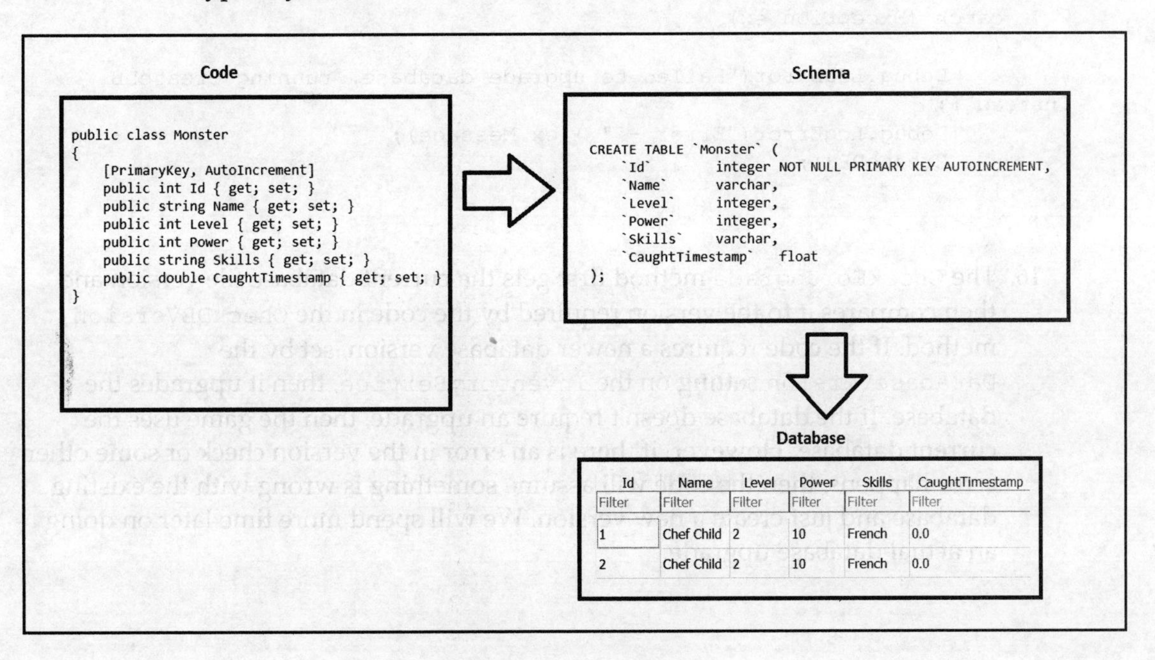

An ORM example of monster to database

14. After the object tables are created, we create a new `DatabaseVersion` object and store it in the database using the `Insert` method on the `_connection`. The `DatabaseVersion` object is very simple and only has one property, called `Version`. We use this object/table to track the version of the database.
15. Remember that, if we don't need to create a new database, then we will check for an upgrade with the `CheckForUpgrade` method, as follows:

```
private void CheckForUpgrade()
{
    try
    {
        var version = GetDatabaseVersion();
        if (CheckDBVersion(version))
        {
            //newer version upgrade required
            Debug.LogFormat("Database current version {0} - upgrading to
{1}", version, DatabaseVersion);
            UpgradeDB();
            Debug.Log("Database upgraded");
        }
    }
    catch (Exception ex)
    {
        Debug.LogError("Failed to upgrade database, running CreateDB
instead");
        Debug.LogError("Error - " + ex.Message);
        CreateDB();
    }
}
```

16. The `CheckForUpgrade` method first gets the current database file version and then compares it to the version required by the code in the `CheckDBVersion` method. If the code requires a newer database version, set by the `DatabaseVersion` setting on the `InventoryService`, then it upgrades the database. If the database doesn't require an upgrade, then the game uses the current database. However, if there is an error in the version check or some other error happens, then the code will assume something is wrong with the existing database and just create a new version. We will spend more time later on doing an actual database upgrade.

17. Finally, we will review the `CreateMonster` method called by the `CatchSceneController`:

```
public Monster CreateMonster()
{
    var m = new Monster
    {
        Name = "Chef Child",
        Level = 2,
        Power = 10,
        Skills = "French"
    };
    _connection.Insert(m);
    return m;
}
```

18. The `CreateMonster` method currently just creates a hardcoded `Monster` object and inserts it into the database using the `_connection.Insert` method. It then returns the new object to the calling code. If you have experience working with relational databases and writing **SQL** code, hopefully, you can appreciate the simplicity of the `Insert` here. We will update the `CreateMonster` and other operation methods, in the next section of this chapter.

Monster CRUD operations

Currently, our Inventory service only creates the same monster, and we need it to create new monsters and perform other operations, such as read, update, and delete monsters. By the way, the standard database operations of Create, Read, Update, and Delete is often referred to as CRUD. So, roll up your sleeves; we are actually going to do a little coding and build the monster CRUD.

Open up the `InventoryService` script in the editor of your choice again and scroll down to the `CreateMonster` method. Delete the `CreateMonster` method and perform the following instructions to replace it and add the other new methods:

- `CREATE`: Add the following method in order to replace the `CreateMonsters` method:

```
public Monster CreateMonster(Monster m)
{
    var id = _connection.Insert(m);
    m.Id = id;
    return m;
```

```
}
```

Instead of creating a hardcoded monster, we now take a monster object and insert it as a new object/record in the database. This returns the new auto incremented id, which we set on the `Monster` object, and return it back to the calling code. We will let another part of our code create the details of the monster.

- `READ` (single): We will handle two versions of the read method: one to read or find an individual monster and another to read all the monsters. Add the following code to read a single monster:

```
public Monster ReadMonster(int id)
{
 return _connection.Table<Monster>()
 .Where(m => m.Id == id).FirstOrDefault();
 }
```

This method takes an id and finds the matching object monster in the table using the `Where` method, which takes a function delegate as a parameter. The code looks like it is using Linq to SQL, but it is not. The `Where` and `FirstOrDefault` are added as part of the SQLite implementation to remain cross-platform.

iOS does not currently support Linq, which often causes confusion for developers coming from a traditional C# background even on other platforms, such as Linux or Mac.
If you want your application to be cross-platform compatible, avoid using the System.Linq namespace entirely.

- `READ` (all): Handling the reading of all monsters is even simpler:

```
public IEnumerable<Monster> ReadMonsters()
{
    return _connection.Table<Monster>();
}
```

One line of code is all it takes to pull all the monsters from the database. It doesn't get much simpler than that.

- `UPDATE`: Update the monster code as follows:

```
public int UpdateMonster(Monster m)
{
 return _connection.Update(m);
```

```
}
```

The `UpdateMonster` method takes a monster object and updates it in the database. An `int` value containing the number of records/monsters updated will be returned. This should always return 1. Note that the monster object passed to the update method should have an existing ID. If the monster object's `Id` property is `0`, then you should instead use the `CreateMonster` method instead. The `CreateMonster` method will create a new monster in the database and set the `Id` property.

- `DELETE`: Finally, when we have no more use for a monster object, we would want to be able to delete it, using the following code:

```
public int DeleteMonster(Monster m)
{
 return _connection.Delete(m);
 }
```

The `DeleteMonster` method is like the `UpdateMonster` method. It takes the monster you want to delete and then deletes it from the database. Returning the number of objects it deleted, which should only be 1. Again, the monster object must have a valid `Id` property. If it doesn't have a valid `Id`, then it doesn't really exist in the database anyway.

Hopefully, you appreciate the ease with which we coded those basic monster CRUD operations. Having the object relational mapping available as part of the SQLite4Unity3d wrapper allowed us to quickly implement the database persistence for our monster inventory. At no point, we even have to utter the words SQL, never mind write any SQL code. Furthermore, in the future, implementing other objects in the Inventory service should be just as easy.

Updating the Catch scene

When we implemented the new CRUD operations for storing our monsters in the Inventory, we broke the existing `CatchSceneController` script. If you recall, we deleted the old sample `CreateMonster` method and wrote a new method to just create a monster entry in the database. This means that not only do we need to fix our updated code, but we also need a way to instantiate new random monster properties.

As usual, before we fix the scene controller, let's address the matter of creating new random monster properties. The ideal solution here is to create a simple static class called a **MonsterFactory** that will randomly build the monsters. Follow the directions to build our new `MonsterFactory` script:

1. Right-click (press *Ctrl* and click on a Mac) the `Assets/FoodyGo/Scripts/Services` folder in the **Project** window. From the context menu, select **Create | C# Script**. Rename the script `MonsterFactory`.
2. Double-click on the new script to open it in the editor of your choice.
3. Edit the file so that the script matches the following:

```
using packt.FoodyGO.Database;
using UnityEngine;

namespace packt.FoodyGO.Services
{
    public static class MonsterFactory
    {
    }
}
```

4. We virtually stripped out the start-up script because we just want a simple `static class`.
5. Next, we will add a list of random names that we will conjugate into a monster name. Add the following field just inside the class:

```
public static class MonsterFactory
    {
        public static string[] names = {
            "Chef",
            "Child",
            "Sous",
            "Poulet",
            "Duck",
            "Dish",
            "Sauce",
            "Bacon",
            "Benedict",
            "Beef",
            "Sage"
            };
```

6. Feel free to add as many other names as you want to the list; just remember that you should follow every entry with a comma, except for the last one.

7. After this, we will create a few other fields to hold skills and other maximum properties:

```
public static string[] skills =
        {
            "French",
            "Chinese",
            "Sushi",
            "Breakfast",
            "Hamburger",
            "Indian",
            "BBQ",
            "Mexican",
            "Cajun",
            "Thai",
            "Italian",
            "Fish",
            "Beef",
            "Bacon",
            "Hog",
            "Chicken"
        };

        public static int power = 10;
        public static int level = 10;
```

8. Just like the names list, feel free to add your own entries to the skills list, as well. Keep in mind that a skill is like a specialty in a cuisine or food product. We will use the skills later in the game when we send our monsters to restaurants in order to find jobs.
9. With the properties out of the way, we will now get down to write the `CreateRandomMonster` method and helpers:

```
public static Monster CreateRandomMonster()
{
    var monster = new Monster
    {
        Name = BuildName(),
        Skills = BuildSkills(),
        Power = Random.Range(1, power),
        Level = Random.Range(1, level)
    };
    return monster;
}

private static string BuildSkills()
{
```

```
    var max = skills.Length - 1;
    return skills[Random.Range(0, max)] + "," + skills[Random.Range(0,
max)];
}
private static string BuildName()
{
    var max = names.Length - 1;
    return names[Random.Range(0, max)] + " " + names[Random.Range(0, max)];
}
```

10. The code is fairly straightforward, but we will review a few things. We use `Random.Range` in the main method and helpers (`BuildName`, `BuildSkills`) to provide a random range of values. For the name and skills helper methods, those random values are used as an index into the `names` or `skills` arrays to return a string value. The random values are then combined into either a name- or comma-separated set of skills.
11. The `Power` and `Level` properties are easily set again using the `Random.Range` method. Use a value of `1`, the start, to the maximum property value we set above.
12. As always, when you are done with editing the script, save it.
13. Open up the `CatchSceneController` script in the Unity editor from the `Assets/FoodyGo/Scripts/Controllers` folder.

Depending on your code editor, it may be difficult to find a file in your script editor. This is why we always reference back to the Unity editor when opening new scripts.

14. Select and delete the `Start` method at the top of the file. We will be rewriting the code, as follows:

```
void Start()
{
    var m = MonsterFactory.CreateRandomMonster();
    print(m);
}
```

15. These couple of lines of code mean that, when the `CatchScene` becomes initialized, a new, random monster will be generated.

16. After you are done with editing, save the file and return to the Unity editor. Wait for the scripts to compile and then press Play, to start the scene. Check out the **Console** window for the output of the new monster properties, and you will see more random values, as shown in the following sample:

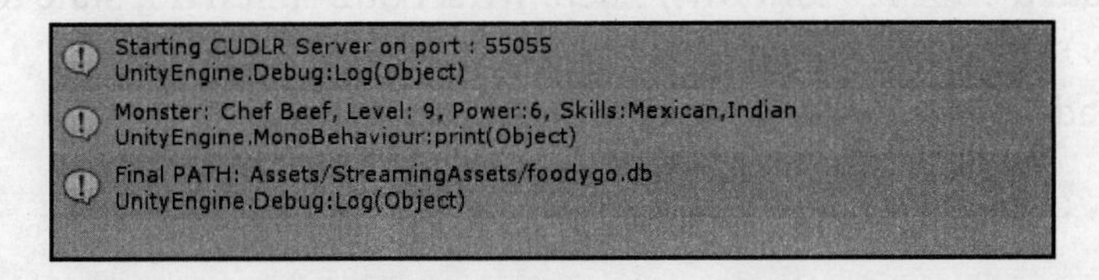

Example of randomly generated monster properties in the Console

So, now when the `CatchScene` starts, random monsters will be created for the player to catch. However, we want the monsters attributes to determine how easy or difficult it is for the player to catch them. What we need to do is add some code that makes the monster more difficult to catch and a way for them to escape. Perform the following to add some difficulty to the `CatchScene`:

1. The first thing we will do is add some new fields to the `CatchSceneController` script. Open the `CatchSceneController` script from the `Assets/FoodyGo/Scripts/Controllers` folder.
2. After the existing fields and just above the `Awake` method, add the new fields:

```
public Transform escapeParticlePrefab;
public float monsterChanceEscape;
public float monsterWarmRate;
public bool catching;
public Monster monsterProps;
```

3. `escapeParticlePrefab` will be our particle effect when the monster escapes.`monsterChanceEscape` determines the chance of escape. `monsterWarmRate` sets how quickly the monster warms after being hit. `catching` is just a `bool` variable we will use to exit from a loop. Finally, `monsterProps` stores the randomly generated monster properties.
4. Change the code in the `Awake` method, to the following code:

```
monsterProps = MonsterFactory.CreateRandomMonster();
print(monsterProps);

 monsterChanceEscape = monsterProps.Power * monsterProps.Level;
 monsterWarmRate = .0001f * monsterProps.Power;
 catching = true;
 StartCoroutine(CheckEscape());
```

5. First, we store the new random monster properties in a variable called `monsterProps`. Then, as you can see by the code, we derive the escape chance by multiplying the monster's power and level together. After we modify that warm rate by multiplying a base value by the power. (Don't worry too much about the hardcoded `.0001f` for now.) Then, we set our `catching` state to `true` and finally, start a coroutine called `CheckEscape`.
6. Now, add the `CheckEscape` coroutine, as follows:

```
IEnumerator CheckEscape()
{
    while (catching)
    {
       yield return new WaitForSeconds(30);
       if (Random.Range(0, 100) < monsterChanceEscape && monster!= null)
       {
           catching = false;
           print("Monster ESCAPED");
           monster.gameObject.SetActive(false);
           Instantiate(escapeParticlePrefab);
           foreach (var g in frozenDisableList)
           {
               g.SetActive(false);
           }
       }
    }
}
```

7. Inside the `CheckEscape` coroutine, there is a while loop that will keep running as long as `catching` is `true`. The first statement inside the loop yields or effectively pauses the loop for `30` seconds, which means the contents of the `while` loop will only run every `30` seconds. After that pause, there is a test to check whether a random value from `0–100` is less than our `monsterChanceEscape`. If it is, and the monster (`MonsterController`) is not null, the monster escapes.
8. When the monster escapes, a few things happen. First, we set the `catching` state to `false`, which stops the loop. Next, we print a message to the **Console**, always a good practice. After this, we disable the `monster.gameobject`, instantiate that escape particle, and finally, disable the scene items. In order to disable the scene items, we iterate over the `frozenDisableList`.
9. Just inside the if statement of the `OnMonsterHit` method, enter the new line of code highlighted:

```
monster = go.GetComponent<MonsterController>();
```

```
if (monster != null)
{
   monster.monsterWarmRate = monsterWarmRate;
```

10. This line just updates the monster (`MonsterController`) `monsterWarmRate` to the same value we calculated above in the `Awake` method.
11. While still in the `OnMonsterHit` method, add a couple more lines just after the `print("Monster FROZEN");` statement, as highlighted:

```
print("Monster FROZEN");
//save the monster in the player inventory
InventoryService.Instance.CreateMonster(monsterProps);
```

12. This line of code just saves the monster properties or monster object to the database inventory after they have been caught. Remember, we need to use the `CreateMonster` method when we are adding a new monster to the inventory.
13. After you have done editing, save the file and return to the Unity editor. Ensure that the code compiles without issue.
14. Before we can test the changes, we need to create and add a new `escapeParticlePrefab` to the `CatchSceneController`.
15. From the `Assets/Elementals/Prebabs(Mobile)/Fire` folder in the **Project** window, drag the `Explosion` prefab into the **Hierarchy** window and drop it. You will see the explosion particle effect play in the scene.
16. Select the `Explosion` object in the **Hierarchy** window. Rename the object to **EscapePrefab** and set the **Transform** component **Z** position to `-3`.
17. Now, drag the `EscapePrefab` object into the `Assets/FoodyGo/Prefabs` folder to create a new prefab. The reason we created a new prefab is because we changed the default position of the object relative to our scene.
18. Delete the `EscapePrefab` from the scene by selecting it in the **Hierarchy** window and pressing the d*elete* key.
19. Select the `CatchScene` object in the **Hierarchy** window. Drag the `EscapePrefab` from the `Assets/FoodyGo/Prefabs` folder onto the empty **Escape Particle Prefab** slot on the **Catch Scene Controller** component in the **Inspector** window.
20. Save the scene and project. Press Play in the editor window to run and test game. Try playing the scene a few times and notice the difference in how easy or difficult it is to catch a monster now. Notice that monsters at the higher-power levels are now impossible to catch. Not to worry, we will offset that later when we add different freeze level balls to a new inventory section in a future chapter.

21. Of course, build and deploy the game to your mobile device and test. As it is right now, it appears the scene just hangs when the monster escapes or is caught. We will fix that later within the `GameManager` when we wire everything together.

Thus far, we now have the `MonsterFactory` generating us random monsters. The monster attributes set the difficulty level on the Catch scene. Then, when the monster is caught, we store its attributes in the new `InventoryService` within the SQLite database. Sounds like our next task is to build the UI for the Inventory, which we will get to in the next section.

Creating the Inventory scene

The great thing about dividing the game content into scenes is our ability to develop and test each functional piece in isolation. We don't have to worry about other game start-up, events, or functionality slowing down our development. Yet, at some point, we need to bring all those pieces together, and it is important to frequently test the whole game, as well.

Before we start working on the Inventory scene, we will do another full reset import of all scripts, as we have done in the previous chapters. This means we will be importing several new and updated scripts and won't be able to cover the extensive changes in detail. We won't have time in the remaining chapter to highlight any interesting code but it is recommended that you wander through the code at your leisure. For those who have made your own changes to the scripts, it is recommended that you back those items up on your own. Follow the directions here to do the asset import:

1. Save your scene and project and do a backup of the project to another location if you made any changes you want to preserve.
2. From the menu, select **Assets** | **Import Package** | **Custom Package...** to open the **Import package...** dialog.
3. From the dialog, navigate to the book's downloaded source code folder `Chapter_6_Assets` and select the `Chapter6_import2.unitypackage` file. Then, click on **Open** to start importing the assets.
4. When the **Import Unity Package** dialog opens, just make sure that all the items are selected and then click on the **Import** button.

With the updated and new scripts loaded, let's build the new Inventory scene by following these directions:

1. From the menu, select **File** | **New Scene**. This will create a new empty scene with a camera and directional light.
2. From the menu, select **File** | **Save Scene As...**. When prompted, name the scene `Inventory` and save it.
3. From the menu, select **GameObject** | **Create Empty**. Rename the new object `InventoryScene` and reset the transform to zero in the **Inspector** window.
4. Drag the `Main Camera` and `Directional Light` objects onto the new `InventoryScene` object in the **Hierarchy** window. This will reparent the objects,as we did for other scenes.
5. From the menu, select **GameObject** | **UI** | **Panel.** This will add a `Panel` with a `Canvas` parent and an `EventSystem` object.
6. Select `EventSystem` in the **Hierarchy** window and press *delete* to remove it. Remember, Unity will add one for us later.
7. Select the `Canvas` parent object and rename it `InventoryBag` in the **Inspector** window.
8. In the **Hierarchy** window, drag the new `InventoryBag` object onto the `InventoryScene` to make it a child.
9. In the **Hierarchy** window, select the `Panel` object. Then, in the **Inspector** window, change the **Image** component color by clicking on the slot. This will open the **Color** dialog selector. Change the **Hex** value to `#FFFFFF` at the bottom of the dialog and then close it. This will set the entire **Game** window to a white background.

This establishes the base of our `InventoryScene`. Before we immerse ourselves more deeply in some of the components, we will first create the inventory item prefab:

1. Select the `InventoryBag` object in the **Hierarchy** window. Then, from the menu, select **GameObject** | **UI** | **Button**. This will add a new `Button` object as a child of `InventoryBag`.
2. Rename the `Button` to `MonsterInventoryItem`, and set the **Rect Transform** | **Anchor Presets** to **top-stretch** while holding the **pivot** and **position** keys in the **Inspector** window.

3. Remove the **Image** component by clicking on the gear icon beside it and selecting **Remove Component** from the drop-down menu.
4. You will now see a warning message in the **Button** component. Change the **Button** component **Transition** property to none. This will remove the warning.
5. From the `Assets/FoodyGo/Scripts/UI` folder in the **Project** window, drag the `MonsterInventoryItem` script onto the `MonsterInventoryItem` button object in the **Hierarchy** window. This will add the inventory script to the object.
6. Right-click (press *Ctrl* and click on a Mac) the `MonsterInventoryItem` in the **Hierarchy** window. From the context menu, select **UI** | **Raw Image**.
7. With the `Raw Image` object selected in the **Inspector** window, change the **Raw Image** | **Texture** property by clicking on the bullseye icon beside the field. From the **Select Texture** dialog, choose the **monster** texture. Also, change the **Rect Transform** | **width** and **height** properties to a value of `80`.
8. From the **Hierarchy** window, select the child `Text` object of `MonsterInventoryItem` and press *Ctrl* + *D* (*command* + *D*, on a Mac) to duplicate the object.
9. Select the first `Text` object and rename it `TopText` in the **Inspector** window. Also, change the component **Text** | **Paragraph** | **Alignment** to center-top, as shown in the following screenshot:

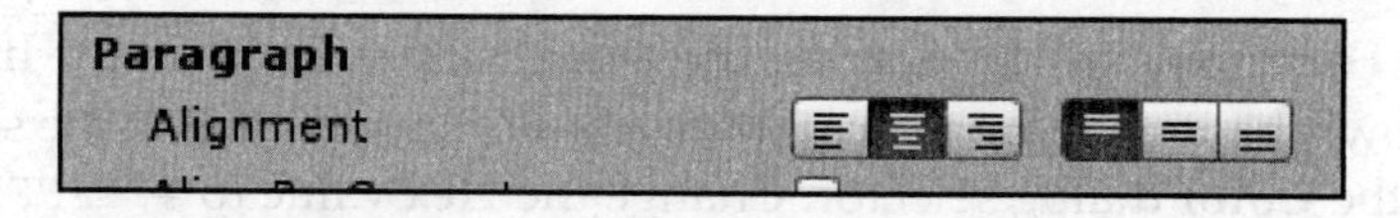

Setting UI text alignment to center-top

10. Repeat the process for the second `Text (1)` object, but rename the object `BottomText` and set its paragraph alignment to center-bottom.
11. In the **Hierarchy** window, drag the `Raw Image` object to just below the `MonsterInventoryItem` so that it is the first child.
12. From the **Hierarchy** window, drag the `MonsterInventoryItem` into the `Assets/FoodyGo/Prefabs` folder in the **Project** window. This will make the `MonsterInventoryItem` into a prefab. Keep the original object in the scene.

With our inventory item created, we will go back and finish the inventory bag:

1. Select the `InventoryBag` object in the **Hierarchy** window, and then from the menu, select **GameObject | UI | Scroll View**. This will add a `Scroll View` beside the `Panel`. Drag the `Scroll View` onto the `Panel` to make it a child, as shown in the following screenshot:

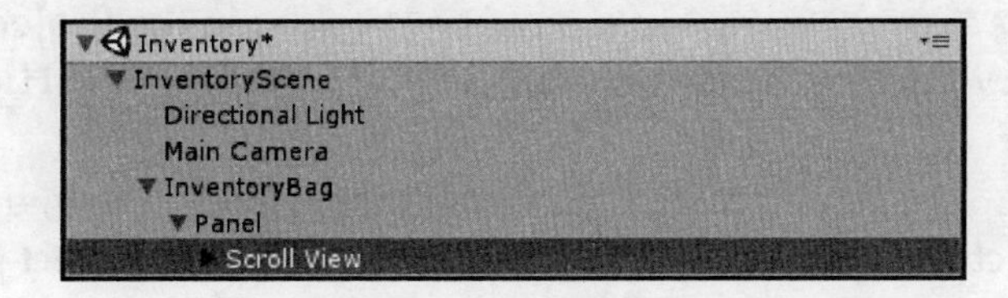

InventoryScene hierarchy thus far

2. Select the `Scroll View` object in the **Hierarchy** window. Then, in the **Inspector** window, change the **Rect Transform | Anchor Presets** to **stretch – stretch** while holding down the **pivot** and **position** keys, as follows:

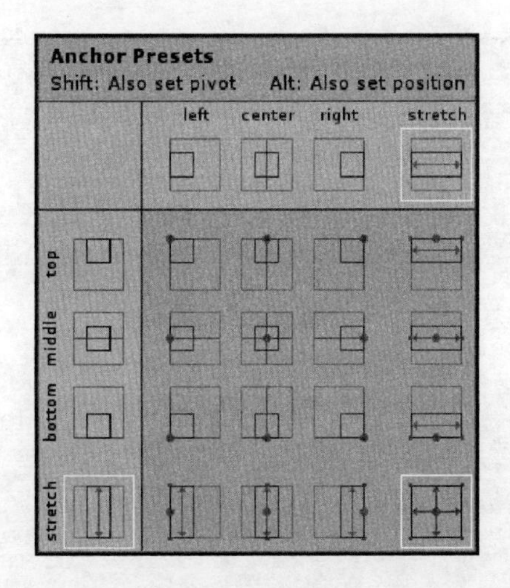

Selecting anchor preset stretch-stretch while holding the pivot and position keys

3. With the `Scroll View` still selected, uncheck the **Horizontal** scroll option from the **Scroll Rect** component. We only want the inventory to be vertically scrollable.
4. In the **Hierarchy** window, expand the `Scroll View` object and then expand the child `Viewport` object as well. This will expose a bottom-level object called `Content`. Select the `Content` object.

5. In the **Inspector** window, set the **Rect Transform** | **Anchor Presets** to **top – stretch** again by holding down the **pivot** and **position** keys.
6. With the `Content` object selected in the **Hierarchy** window from the menu select **Component** | **Layout** | **Grid Layout Group** to add the component to the `Content` object. Repeat the process by selecting **Component** | **Layout** | **Content Size Fitter**.
7. From the `Assets/FoodyGo/Scripts/UI` folder in the **Project** window, drag the `InventoryContent` script onto the `Content` object in the **Hierarchy** or **Inspector** windows.
8. With the `Content` object selected in the **Inspector** window, drag the `Scroll View` object onto the empty **Inventory Content** | **Scroll Rect** property.
9. Then, from the `Assets/FoodyGo/Prefabs` folder in the **Project** window, drag the `MonsterInventoryItem` prefab onto the empty **Inventory Content** | **Inventory Prefab** slot.
10. Confirm or set the values for **Rect Transform**, **Grid Layout Group**, **Content Size Fitter** and **Inventory Content** to the values shown in the following screenshot:

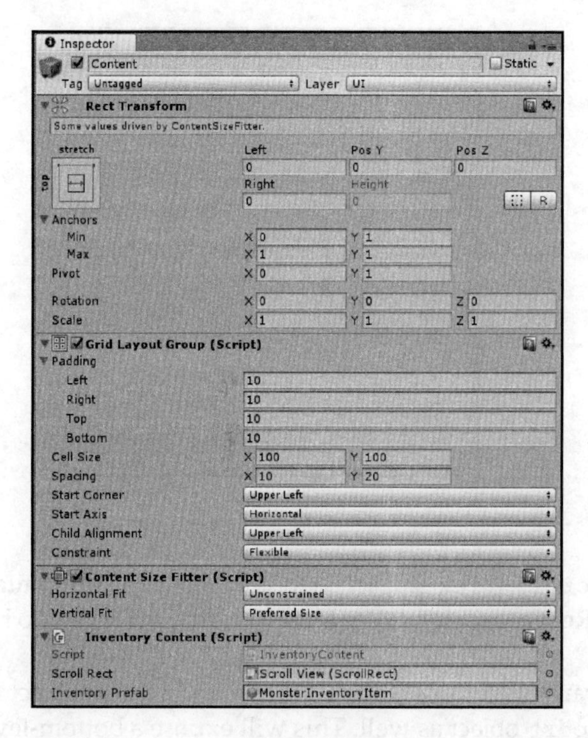

Content object configuration

11. Finally, in the **Hierarchy** window, drag the `MonsterInventoryItem` onto the `Content` object, to make it a child. Then, deactivate the object by unchecking the checkbox beside the object name in the **Inspector** window. We will just use that object as a reference. Your updated Hierarchy window should match the following screenshot:

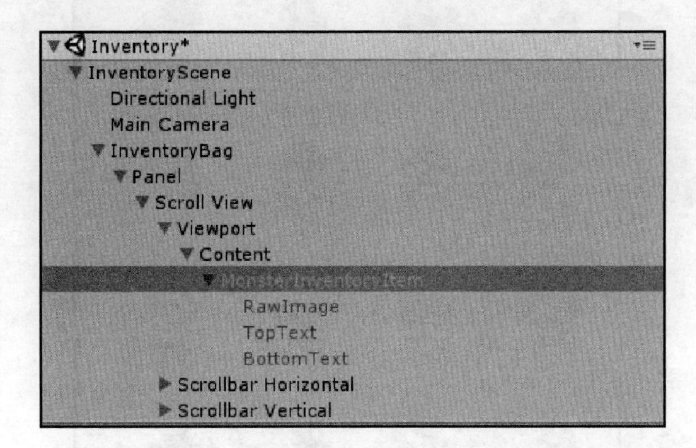

Inventory hierarchy

Thus far, we have most of the **Inventory** scene constructed and now all we have to do is connect things together. Perform the following instructions to finish the scene:

1. Drag `InventorySceneController` script from the `Assets/FoodyGo/Scripts/Controllers` folder in the **Project** window and drop it onto the `InventoryScene` object in the **Hierarchy** window.
2. Select the `InventoryScene` object. Drag the `Content` object from the **Hierarchy** window onto the empty **Inventory Scene Controller** | **Inventory Content** slot in the **Inspector** window.
3. From the `Assets` folder in the **Project** window, drag the **Catch** scene onto the **Hierarchy** window. This allows us to see both scenes overlapping each other.
4. Select and drag the `Services` object from the `Catch` scene and drop it onto the **Inventory** scene. This will add the `Services` to the **Inventory** scene. If you recall, we only used those services for testing and planned to remove them from the Catch scene later, anyway.
5. Right-click (press *Ctrl* and click, on a Mac) on the **Catch** scene to open the context menu. Select **Remove Scene** and when prompted to save, click on the **Save** button.

6. Press **Play** to run the scene and see the results. The following is a sample screenshot of several captured practice monsters caught during play testing the Catch scene:

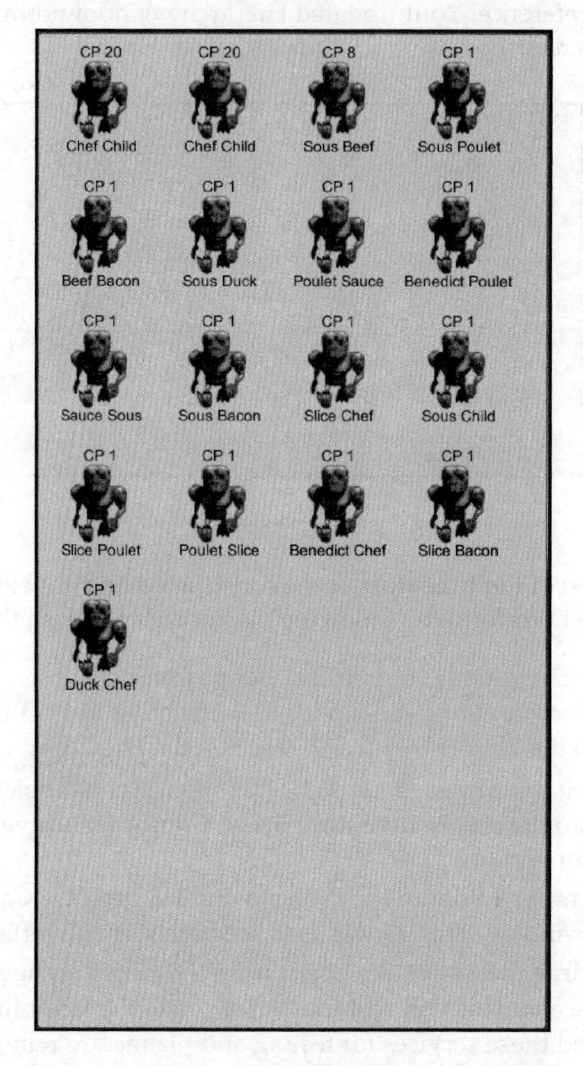

Inventory scene showing the caught monsters

The 3D character is designed by Reallusion iClone Character Creator. To create more
customized characters, please visit `http://www.reallusion.com/iclone/character-creator/default.html`for more details.

Hopefully, you spent some time catching monsters in the test **Catch** scene after we connected the Inventory service and can now see those caught monsters. If you have no monsters showing, not to worry we will wire the game together before the end of the chapter. Also, you may have noticed that our monster inventory items are actually buttons, but they don't do anything. That is also fine, we will add the detail section of the inventory later. For now though, we want to finish up the chapter by connecting all the scenes together in a proper game.

Adding the menu buttons

In order to connect our scenes together, we need player input to trigger an event. We do have an event setup for when the player clicks on the monster in the Map scene. However, we also want the player to move from the Map/Catch scene to the Inventory scene and back. In order to do this, we will add some UI buttons to each of the scenes.

Since we already have the **Inventory** scene already open, let's begin by adding the new button to that scene:

1. Right-click (press *Ctrl* and click, on a Mac) on the `InventoryBag` object in the **Hierarchy** window and select **UI** | **Button** from the context menu. This will add a button just below the `Panel`. Expand the `Button` object and then select the child `Text` object and press *delete* to remove it.
2. Select the new `Button` object and rename it `ExitButton` in the **Inspector** window.
3. From the **Inspector** window, set the **Rect Transform** | **Anchor Presets** to **bottom-center** while holding the **pivot** and **position** keys. Then, change the **Rect Transform** properties' **Width** and **Height** to 75 and the **Pos Y** to 10.
4. Change the **Image** | **Source Image** by clicking on the bullseye icon beside the slot to open the **Select Sprite** dialog. Choose the sprite called `button_set11_b` from the dialog.
5. Finally, we will wire up the button by clicking on the **+** under the **Button** | **On Click** event property. This will create a new event slot. Then, drag the **InventoryScene** object from the **Hierarchy** window onto the empty **None(object)** slot. From the **No Function** drop-down menu, select **InventorySceneController** | **OnCloseInventory**.

6. The full configuration for the `ExitButton` is shown in the following screenshot:

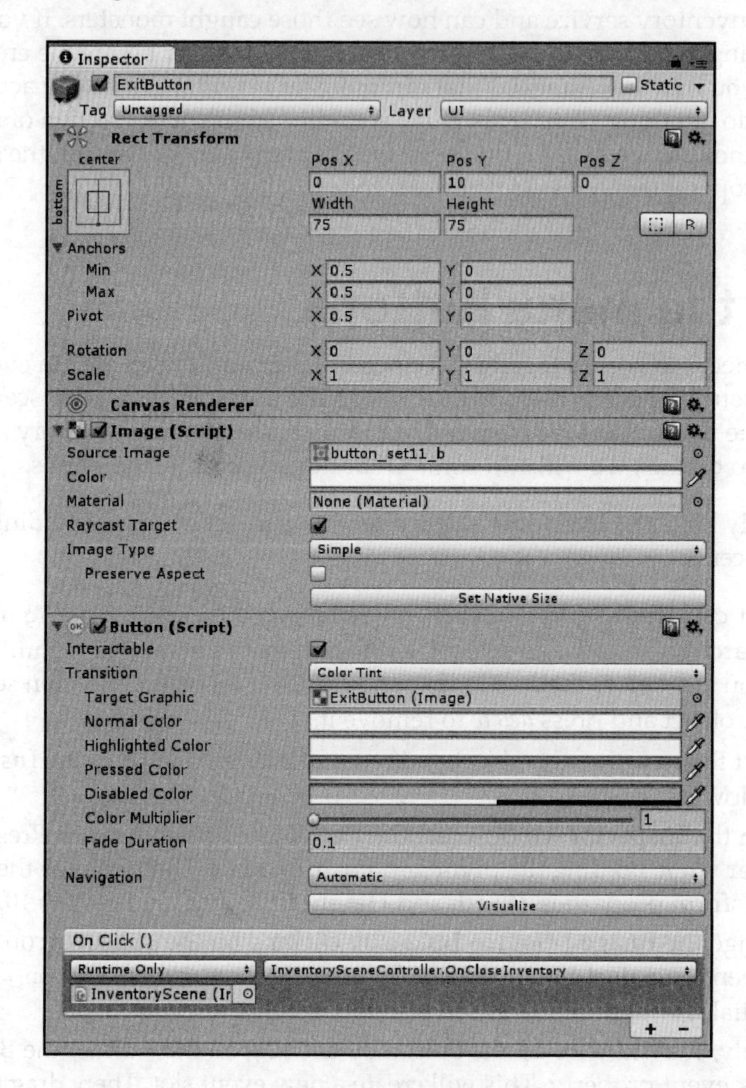

ExitButton component configuration

The `ExitButton` we just added is for closing the **Inventory** scene and returning to the scene that opened it. Before we get the **Map** and **Catch** scenes, we first need to transfer our services to the **Game** scene and take care of some minor updates. Perform the instructions to update the **Game** scene:

1. From the `Assets` folder in the **Project** window, drag the **Game** scene to the **Hierarchy** window.
2. Drag the `Services` object from the **Inventory** scene to the **Game** scene. This will be the new permanent home for the `Services` object. Rename the `Services` to `_Services` in the **Inspector** window. Again, we use the _ to denote objects that are not to be destroyed.
3. Remove the **Inventory** scene by right-clicking (press *Ctrl* and click, on a Mac) on it and selecting **Remove Scene** from the context menu. When prompted, save the scene by clicking on **Save**.
4. Select the `_GameManager` object in the **Hierarchy** window, and update the **Game Manager** script scene names, as shown in the screenshot:

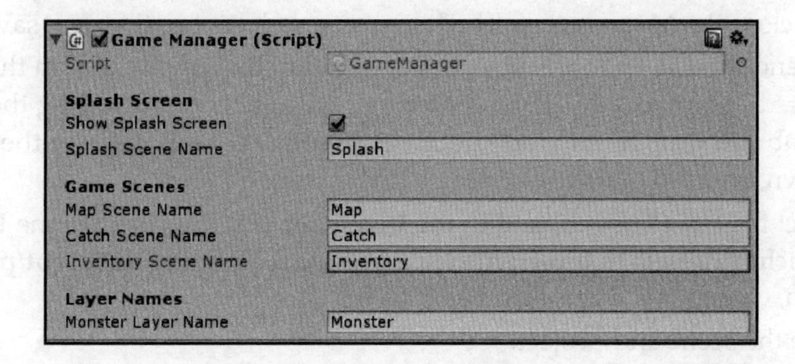

Game Manager Scene names updated

Next, we will add a new button to access the **Inventory** scene on the **Map** scene. Follow the instructions to add the new button:

1. Double-click on the **Map** scene in the `Assets` folder in the `Project` window. This will close the **Game** scene, and when prompted, ensure that you save the changes.
2. Expand the `MapScene` object in the **Hierarchy** window, and right-click (press *Ctrl* and click, on a Mac) to open the context menu. From the menu, select **UI | Button**, to add a new button.

3. Select the button created in previous step, and in the **Inspector** window, rename it `HomeButton`. Set the **Rect Transform** | **Anchor Presets** to **bottom-center** while holding the **position/pivot** keys. Change the **Rect Transform** | **Width/Height** to `80` and the **Pos Y** to `10`. Then, change **Image** | **Source Image** to `button_set06_b` by clicking on the bullseye icon and selecting the sprite for the **Select Sprite** dialog.
4. Drag the `HomeButton` script from the `Assets/FoodyGo/Scripts/UI` folder in the **Project** window and drop it onto the `HomeButton` object in the **Hierarchy** window.
5. Select and drag the `HomeButton` object from the **Hierarchy** window to the `Assets/FoodyGo/Prefabs` folder in the **Project** window to make it a new prefab.

Finally, the last scene we need to do is the **Catch** scene, which will make our round trip complete. Perform the following directions to add the `HomeButton` to the scene:

1. Double-click on the **Catch** scene in the `Assets` folder of the **Project** window. This will close the **Map** scene, and when prompted, ensure that you save the changes.
2. Expand the `CatchScene` object in the **Hierarchy** window. From the `Assets/FoodyGo/Prefabs` folder in the **Project** window, drag the `HomeButton` prefab and drop it onto the `Catch_UI` object. You will now see the `HomeButton` drawn over top of the `CatchBall` in the scene.
3. Select the `HomeButton`, and in the **Inspector** window, change the **Rect Transform** | **Anchor Presets** to **top-right** while, of course, holding the **pivot/position** keys. Then, change the **Pos X** and **Pos Y** to `-10`.
4. Save the scene and project.

Now, with the scene transition buttons added, all the scripts updated, and everything else set, it is time to bring everything together and run the game.

Bringing the game together

Wow, now that our game consists of five scenes, it is time to bring everything together in a full game. There are just a couple of things we need to do in order to get everything connected. Perform the following instructions to configure the build settings and test the game:

1. Open the **Game** scene within the editor; this will be our starting scene now.

2. From the menu, select **File** | **Build Settings** to open the **Build Settings** dialog. Drag the scenes from the `Assets` folder in the **Project** window to **Scenes in Build** area. Reorder the scenes by dragging them around so that they match this screenshot:

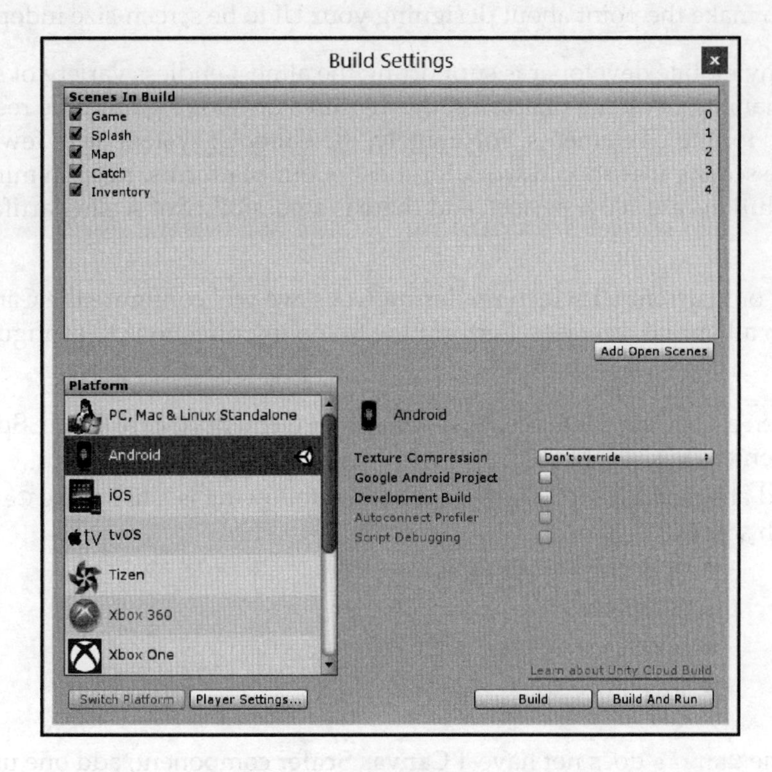

Scenes in build

3. Build and deploy the game to your mobile device and test it. Play the game and try catching monsters, checking the inventory, and so on.

The game runs well, and there are a few bugs, but the one thing you may immediately notice is that the buttons and inventory items are not the same size we designed for. Not to worry, we will deal with the UI scaling issue in the next section. We will fix some of the other bugs you may have noticed as we move through the other chapters. We will also devote an entire section to stabilizing code in `Chapter 9`, *Finishing the Game*.

Mobile development woes

If you were paying attention, you may have noticed early on that our Splash scene text wasn't sizing correctly when deployed to the mobile device. The reason we left that issue until now is to make the point about designing your UI to be screen-size independent.

The bane of any mobile developer is supporting the almost endless variety of screen sizes in a consistent manner. On some platforms, this requires developing multiple resolution images/sprites for the UI elements. Fortunately, the Unity UI system has a few nice options for screen-size scaling that should work for most of our platforms. Keep in mind, however, that not all solutions are 100% perfect, and there is a possibility of scaling artifacts on some platforms.

In order to fix our current UI screen rendering woes, we will configure the Canvas Scaler component on all our UI canvases. Perform the following directions to configure the Canvas Scaler:

1. Open any of the game scenes that have UI elements (Map, Catch, Splash, and Inventory).
2. Find and select the `Canvas` element. The following is a list of the `Canvas` items in each scene:
    - Splash scene: `Canvas`
    - Map scene: `UI_Input`
    - Catch scene: `Catch_UI`, `Caught_UI`
    - Inventory scene: `InventoryBag`
3. If the `Canvas` does not have a **Canvas Scaler** component, add one using the **Add Component** button.
4. Set all the **Canvas Scaler** properties to the following:

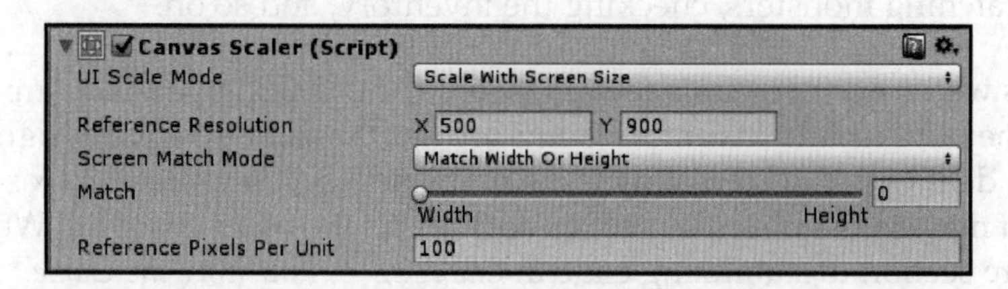

Canvas Scaler component properties

5. Repeat this process for all scenes and `Canvas` objects. Remember to save the scenes after you make the changes.

6. Build and deploy the game to your mobile after you have made all the changes. Test the game and notice how the UI elements now scale as they were designed to.

Summary

In this chapter, we started out slowly by first understanding how to store the player's caught monsters. After that, we looked at a couple of database options before deciding on a cross-platform object relational mapping tool for SQLite, called SQLite4Unity3d. We wrapped the database into our new Inventory service and then wrote CRUD operations for our monster inventory items. From that, we determined we needed a better way to randomly generate monsters, so we developed a monster factory. That allowed us to loop back and complete the **Catch** scene by incorporating the monster factory for generating monsters and inventory service for storing caught monsters. With monsters being stored in the database, we then developed a new inventory scene to view caught monsters. Finally, we tied everything together with UI menu buttons and joined all the scenes together as a full game. Of course, the chapter ended by resolving some platform deployment issues.

For the next chapter, we will get back to exploring the AR world around the player and the map. We will extend the objects and places the player can interact with on the map through the use of some clever services. That will also require us to go to the next level in our GIS/GPS knowledge to explore spatial queries and other advanced concepts.

7
Creating the AR World

By far, our game has focused on the player interaction with the experimental cooking monsters. Although they can track the monsters on a map, nothing else around them is part of this alternate reality. Well, of course, we want to fix this shortcoming. We want to create a rich AR world around the player that they can also interact with, to find/restock items and train monsters and send them on cooking missions. In order to do that, we need to populate the world around the player with alternate game reality locations.

In this chapter, we will jump back to the map and start to populate the world around the player with new alternate reality locations. These locations won't be entirely based on our game reality. In fact, we will use real-world locations as a basis for our alternate game world. The locations we populate in the AR world will be drawn from a location-based web service. The following is a summary of what we will cover in this chapter:

- Getting back to the map
- The Singleton
- Introducing the Google Places API
- Using JSON
- Setting up the Google Places API Service
- Creating the markers
- Optimizing the search

This will be a short but intense chapter, and we will cover a number of things quickly. If you have jumped ahead from a previous chapter in the book, ensure that you either are an experienced Unity developer or just browsing through the books content. As always, if you have jumped ahead from a previous chapter in the book, open the project in the `Chapter_7_Start` folder from the downloaded book source.

Getting back to the map

Of course, at some point, you had to realize we would come back to the map. The map is a core and fundamental element to our location-based game. It provides the window to the alternate game reality, and yet, still provides a real-world reference for the player. Until now, the only hint we gave to this alternate world was by tracking and visualizing the monsters. However, our monsters are entirely random and do not adhere to any of the real world around them. In some way, this tarnishes the experience, but fixing the monster's behavior could be another book itself. In order to bring some real-world basis back to our virtual objects/places, we will use the Google Maps API to populate the map around the player.

Before we jump into adding new features to the game, let's spend some time fixing a couple issues we ignored at the end of the last chapter. Chances are, you noticed the problem if you played the game for a while. If you didn't though, the issue was that the GPS service would stop running and the map would stop updating after a user returned from the Catch or Inventory scenes. In the previous chapter, we ran out of time to fix this issue, and besides, we knew that in the next chapter we would have the opportunity to fix the problem.

If you are a professional developer, you know all too well that breaking, refactoring, and fixing code is all about the evolution of software and games. Many new developers spend too much time writing out the perfect piece of code and then doing their best to preserve it. Code is meant to be changed, rewritten, and most certainly deleted. The sooner you realize that, the better mind-set you will have as a developer. Of course, there is a time and place for refactoring, and it certainly isn't the day before you ship a game or product.

In order to fix the GPS issue, the `GPSLocationService` class was rewritten as a `Singleton`. Also, all dependent classes that use that service also needed to be updated. We will first do an import of updated scripts and then move some services around, as follows:

1. Open up Unity to either the project as we left it at the end of the last chapter or from the downloaded source code `Chapter_7_Start` folder.
2. From the menu, select **Assets** | **Import Package** | **Custom Package...**, and when the **Import Package...** dialog opens, navigate to the downloaded source code folder `Chapter_7_Assets`.
3. Import the package as you have done before.
4. Open the **Game** scene from the `Assets` folder in the **Project** window, by double-clicking on it.
5. Drag the **Map** scene into the **Hierarchy** window from the `Assets` folder in the **Project** window.

6. Expand the **Map** | **MapScene** object in the **Hierarchy** window. Then, expand the `Services` object as well. Also, expand the `_Services` object in the **Game** scene. The **Hierarchy** window should now resemble the following screenshot:

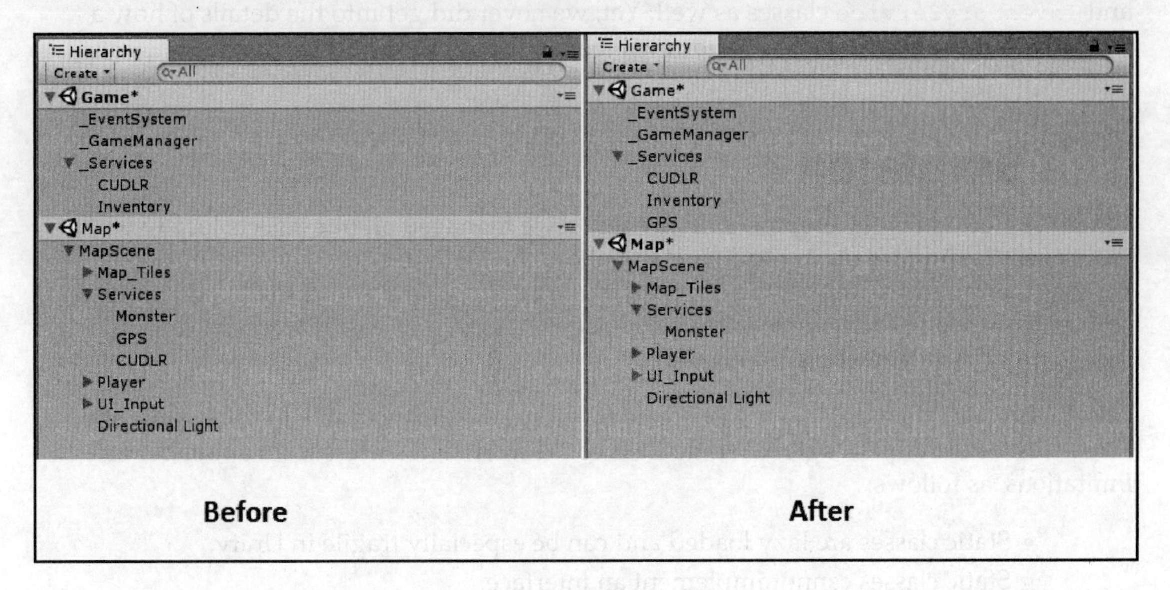

Scenes before and after the update

7. Select the **Map** | **MapScene** | **Services** | **CUDLR** object and press **Delete** to remove it. Our master **Game** scene has CUDLR and we don't need two instances.
8. Drag the **Map** | **MapScene** | **Services** | **GPS** object and drop it onto the **Game** | **Services** object to make it a child. This will promote our `GPS` service to a master service now. The `Monster` service will be the only service left on our **Map** scene, as shown in the preceding screenshot.
9. Right-click (press *Ctrl* and click on a Mac) on the **Map** scene in the **Hierarchy** window to open the context menu. Select **Remove Scene** from the menu and when prompted make sure that you save it.
10. Save the **Game** scene by typing *Ctrl* + *S* (*command* + *S* on a Mac).
11. Press Play and run the game again. Ensure that the GPS simulation mode is running (refer to `Chapter 2`, *Mapping the Player's Location*, if you are unsure how to set that). Change from the **Map** to the **Inventory** scene and back several times. Notice now how the GPS keeps running, as it should.

So, it may appear that all we had to do is move a couple services around, but that was far from the extent of the required changes. As was mentioned, the `GPSLocationService` was converted to a Singleton. We have used the Singleton pattern before for the `GameManager` and `InventoryService` classes as well. Yet, we never did get into the details of how a Singleton works. The Singleton pattern will be our focus in the next section.

The Singleton

When we started the development of our game, we managed all our objects locally within the scene. We didn't want or need to worry about the lifetime of our services/managers. However, as almost always is the case, our game matured, and we started using multiple scenes. Now, we needed our services or manager classes to be easily accessible by child scenes and almost anywhere in our code.

In games of old, we would have just created a global or static variable to track game state across scenes or scripts. A global static class could work, but suffers from a number of limitations, as follows:

- Static classes are lazy loaded and can be especially fragile in Unity.
- Static classes cannot implement an interface.
- Static classes can only be derived from an object. They cannot inherit from `MonoBehaviour` and thus be used as components in Unity, which means that they cannot also use Unity Coroutines or other base methods, such as `Start`, `Update`, and so on.

Let's take a look at the difference in declaring a standard `MonoBehaviour` game object and one that uses `Singleton` with the updated `GPSLocationService` script.

Feel free to open the updated scripts in the editor of your choice while you follow along.

The previous implementation of the `GPSLocationService` was declared as follows:

```
public class GPSLocationService : MonoBehaviour
```

This, as we saw many times now, is the standard way to declare a Unity component. Compare that to the new class declaration of the `GPSLocationService`:

```
public class GPSLocationService : Singleton<GPSLocationService>
```

Review and understand how the `Singleton` class definition works. The script is located in the `Assets/FoodyGo/Scripts/Managers` folder.

This may look strange: declaring an object to inherit from generic type called Singleton with itself as the type. Think of the `Singleton` class as a wrapper that converts the instance to a global static variable that is accessible anywhere in code. Here is an example of how the GPS service was accessed before and is now:

```
//before, GPS service object was set as a field of the class which had to
be set in the editor
public GPSLocationService gpsLocationService;
gpsLocationService.OnMapRedraw += gpsLocationService_OnMapRedraw;

//after, the GPS service is now accessible anywhere as a Singleton
GPSLocationService.Instance.OnMapRedraw +=  GpsLocationService_OnMapRedraw;
```

There is one critical element that you should be aware of when working with our implementation of `Singleton`. We will always make sure that we instantiate a `Singleton` manager or service as a game object in a scene. This is done, so we can take advantage of the `Start`, `Awake`, `Update`, and other methods. However, if you do not add these objects into a scene and then try to directly access them in code, they will be accessible but will likely be missing important initialization set in the `Awake` or `Start` methods.

Take a look at the classes that consume the `GPSLocationService`: the `MonsterService`, `CharacterGPSCompassController`, and `GoogleMapTile`. As you will appreciate, the differences are subtle but significant.

In this chapter, we will add another consumer of the `GPSLocationService` and also make the new service a `Singleton`. This new service will be the `GooglePlacesAPIService`, and we will cover it in the next section.

Introducing the Google Places API

We will use the Google Places API to populate the virtual world around the player with references to real-world locations or places. Since we already used the Google Static Maps API, adding another service should be straightforward. However, unlike the maps API, the places API is far more restrictive on usage. This means that we will need to undertake additional set-up steps and modify the way we access the service. Not to mention, there will be a direct impact on our business model when we ship.

Another direct competitor to the Google Places API is Foursquare. Foursquare has far fewer restrictive usage limits, but does require additional authentication mechanisms. We will revisit this topic again when we will start `Chapter 9`, *Finishing the Game*.

In order to start using the Google Places API, we will need to register and create new API key. This key will allow your app/game to make 1000 queries a day, which isn't very many, when spread over multiple players. Fortunately, if you register for billing with Google, they will bump up your limit to 150,000 requests per day. The code we use in this chapter will attempt to optimize the fewest number of requests so as not to exceed that 1000 request limit in testing.

The Google Static Maps API also has a limit of 2500 requests per IP address. This is far less restrictive and is the reason we didn't register. Furthermore, we only make a request for a new map when the player has moved out of map tile bounds.

Open your favorite web browser, and perform the following instructions to generate a Google Places API key:

1. Click on: `https://developers.google.com/places/web-service/get-api-key` or copy it to your browser.
2. Click on the blue button labeled with **GET A KEY** roughly in the middle of the page, as shown in the following screenshot:

Get an API key
If you are using the standard Google Places API Web Service
To get started using the Google Places API Web Service, click the button below, which guides you through the process of activating the Google Places API Web Service and getting an API key.
GET A KEY

Getting a Google developer key

3. Sign in to your Google account or create if you don't have one.

4. After you sign in, a dialog will prompt you to select or create a new project. Choose create a new project and name it **Foody GO** or whatever else you find appropriate, as shown in the following dialog:

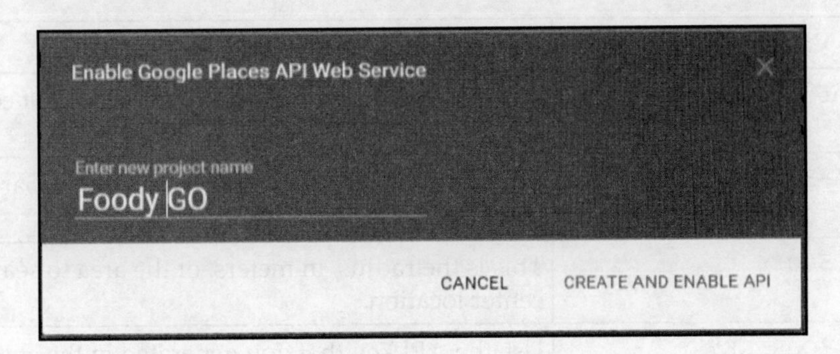

Creating a new Google Places API project

5. Click on the **CREATE AND ENABLE API** link. This will open a dialog showing you your API key and links to starting materials. Ensure that you copy the key as it is shown in the following screenshot:

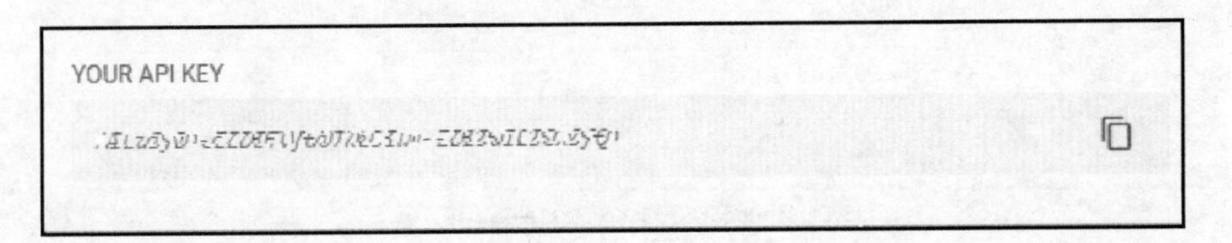

Example key, intentionally obscured

Now that you have your API key, let's test out the REST service that is the Google Places API. This exercise will not only show us what information comes back in a search, but will also help us understand the API. Follow the exercise instructions here:

1. Click on or copy the given URL: `https://www.hurl.it/` to your browser. `Hurl.it` allows you to test REST API calls quickly and easily within your browser window.
2. At the top of the form, enter the Google Places API base URL in the `yourapihere.com` text field, as follows: `https://maps.googleapis.com/maps/api/place/nearbysearch/json`.

3. Next, click on the **Add Parameter** link and fill in the name as `type` and value as `food`. The following is the table defines the parameters and the values that should be used in this exercise:

Name	**Value**	**Description**
`type`	`food`	The type of place you want to search for. Of course, we will use the term food.
`location`	`-33.8670,151.1957`	The latitude and longitude coordinates, separated by a comma.
`radius`	`500`	This is the radius, in meters, of the area to search from the center location.
`key`	`YOUR KEY`	Use the API key that you generated in the above step.

4. Complete the parameters as they are referenced in the table and make sure that they match the following screenshot:

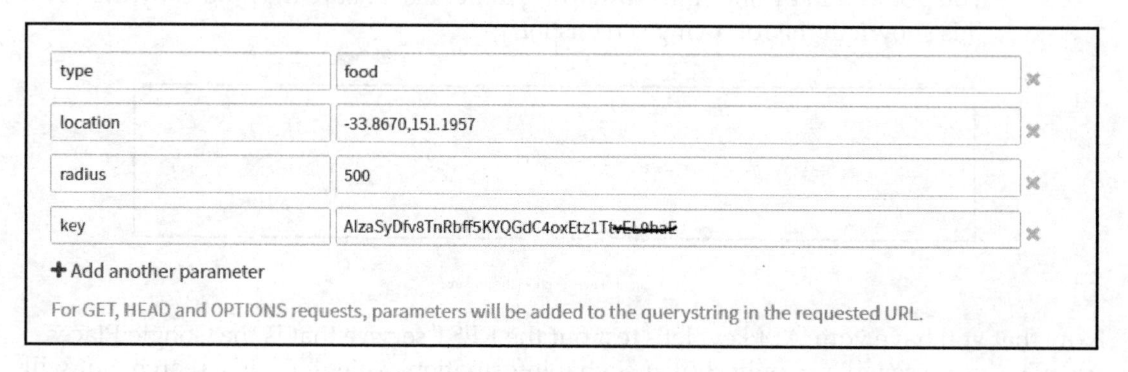

Adding parameters for the nearby search

5. After you are done adding the parameters, confirm that you are not a robot by clicking on the checkbox and then click on **Launch Request**.

6. Assuming you entered the parameters correctly, you should now see a change to: response message below the form. The message will be quite long and may look rather foreign, depending on your previous experience with **JSON**.

Leave your browser open and on the results page. We will come back to explore the output in the next section.

Using JSON

JSON stands for **JavaScript Object Notation** and is a definition for a very lightweight format for transmitting data through object serialization. This means that the message we received from the Google Places API is actually a set of objects. All we need to do is properly parse these objects, and it will be a simple matter to understand the search results. Unity does actually have a built-in JSON library; but, at the time of writing this book, it could not parse the Google Places API response. Fortunately, there are plenty of resources available to parse JSON.

Since the Unity engine could not effectively parse the response, it was decided to use a library called **TinyJson**. TinyJson is another open source library pulled from GitHub, but parts had to be rewritten in order to support the iOS platform. However, a couple calls to the `System.Linq` namespace were left in. If you plan to run this code on an iOS device, make sure that your scripting backend is set to `IL2CPP`.

As we mentioned previously, you have to be careful what C# namespaces you use when developing with iOS in mind. Normally, we try to avoid the System.Linq namespace, as it can be problematic when deploying to iOS.

Now that we have the big picture, we will look at the sample snippet of code that shows how we will make a search request to the API:

```
//this code is run as part of a coroutine
var req = new
WWW("https://maps.googleapis.com/maps/api/place/nearbysearch/json?location=
-33.8670,151.1957&type=food&radius=500&key={yourkeyhere}");

//yield until the service responds
yield return req;
```

```
//extract the JSON from the response
var json = req.text;

//use the TinyJson library JSONParser to de-serialize the result into
//an object called SearchResult
var searchResult = TinyJson.JSONParser.FromJson<SearchResult>(json);
```

This code is very similar to the code we used to download the images from the Google Static Maps API, except now we are using the `WWW` method to return the JSON text string and parsing it into an object called `SearchResult`. `SearchResult` is defined by reading the JSON and extracting the properties and object hierarchy into class definitions. Unfortunately, this has to be a manual process because we can't use any dynamic code generation if we want to support iOS. Fortunately, though, there are plenty of tools available that allow us to convert JSON into the required class definitions.

In order for you to see the complete process and the magic of JSON, we will use an online tool to construct our `SearchResult` class hierarchy. Perform the following instructions to perform the exercise:

1. Go back to the `Hurl.it` page and copy the JSON response. Ensure that you include everything from the starting curly brace (`{`) to the ending curly brace (`}`) at the bottom.
2. Copy the JSON text to your clipboard by typing *Ctrl* + C (*command* + C on a Mac)
3. Open another browser tab to `http://json2csharp.com/`.
4. Paste the JSON you copied earlier into the JSON field by typing *Ctrl* + *V* (*command* + *V* on a Mac).
5. Click on the **Generate** button to create the C# classes, as follows:

```
public class Location
{
    public double lat { get; set; }
    public double lng { get; set; }
}

public class Northeast
{
    public double lat { get; set; }
    public double lng { get; set; }
}

public class Southwest
{
    public double lat { get; set; }
    public double lng { get; set; }
}

public class Viewport
{
    public Northeast northeast { get; set; }
    public Southwest southwest { get; set; }
}

public class Geometry
{
    public Location location { get; set; }
    public Viewport viewport { get; set; }
}

public class OpeningHours
{
    public bool open_now { get; set; }
    public List<object> weekday_text { get; set; }
}

public class Photo
{
    public int height { get; set; }
    public List<string> html_attributions { get; set; }
    public string photo_reference { get; set; }
    public int width { get; set; }
}

public class Result
{
    public Geometry geometry { get; set; }
    public string icon { get; set; }
    public string id { get; set; }
    public string name { get; set; }
    public OpeningHours opening_hours { get; set; }
    public List<Photo> photos { get; set; }
    public string place_id { get; set; }
    public int price_level { get; set; }
    public double rating { get; set; }
    public string reference { get; set; }
    public string scope { get; set; }
    public List<string> types { get; set; }
    public string vicinity { get; set; }
}

public class RootObject
{
    public List<object> html_attributions { get; set; }
    public string next_page_token { get; set; }
    public List<Result> results { get; set; }
    public string status { get; set; }
}
```

Generated class hierarchy from JSON response message

6. We would then copy this code and paste it into our code editor as part of a script and rename the `RootObject` class to be our `SearchResult` class. `RootObject` is just a name assigned to root or top-level unnamed object in the response.

The sample code we have shown above and the generated class hierarchy was used to build the `GooglePlacesAPIService`. The service itself has several other features, but now you at least understand the core of how the service was built and how to build other such services that consume JSON. In the next section, we will set up the new service.

Setting up the Google Places API service

Since we have already imported the updated scripts, setting up this new service should be a breeze for us now. Perform the following instructions to set up and test the `GooglePlacesAPIService`:

1. Go back to the Unity editor. Drag the **Map** scene into the **Hierarchy** window from the `Assets` folder in the **Project** window.
2. Expand the `MapScene` and `Services` object in the **Hierarchy** window.
3. Right-click (press *Ctrl* and click on a Mac) on the `Services` object, and from the context menu select **Create Empty**. Rename the new object `GooglePlacesAPI`.
4. Drag the `GooglePlacesAPIService` script from the `Assets/FoodyGo/Scripts/Services` folder and drop it onto the `GooglePlacesAPI` object in the **Hierarchy** or **Inspector** windows.
5. Right-click (press *Ctrl* and click on a Mac) on the `MapScene` object in the **Hierarchy** window, and from the context menu select **Create Empty**. Rename the new object `PlaceMarker`.
6. Right-click (press *Ctrl* and click on Mac) on the `PlaceMarker` object in the **Hierarchy** window, and from the context menu select **3D Object** | **Cylinder**.
7. Drag the new `PlaceMarker` object into the `Assets/FoodyGo/Prefabs` folder to make it a new prefab. Leave the original object in the scene but deactivate it by unchecking the checkbox beside the objects name in the **Inspector** window.
8. Select the `GooglePlacesAPI` object. Drag the `PlaceMarker` prefab you just created onto the empty slot for the **Place Marker Prefab** slot.

9. With the `GooglePlacesAPI` object still selected, fill in the properties, as shown in the following screenshot:

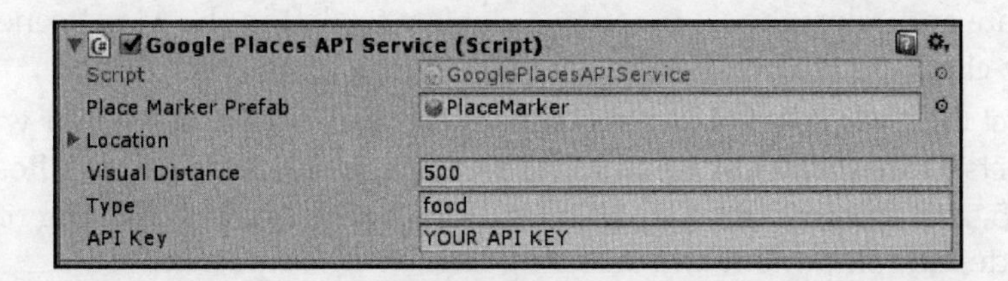

Google Places API Service configuration

10. Ensure that you enter the key you generated for the **API key**, in the previous section.
11. Right-click (press *Ctrl* and click on a Mac) on the Map scene, and from the context menu, select **Remove Scene**. Ensure that you save the scene when prompted.
12. Press **Play** to run the game in the editor and make sure that the GPS service is set to simulate. If you are still using the Google headquarters (37.62814, -122.4265) as your simulation start location, then you should see plenty of place marker cylinder objects spawn around the character.

If you are not using the Google coordinates in your GPS simulation and not seeing any locations, ensure that you try a location close to lots of restaurants, grocery stores, or any other place related to food. Of course, if you continue to experience troubles, refer to `Chapter 10`, *Troubleshooting*.

With the places service running, the player can now see new objects around them. We won't allow the player to interact with those objects, yet. However, we certainly don't want a plain old cylinder representing those markers either. What we need to do is create a much better-looking marker, and we will do that in the next section.

Creating the markers

Generally, when developers prototype a game or game scene, they ignore the details of esthetics and just use some ugly marker. That typically works until something better is provided by the design team. Well, since we don't have a design team now, we will instead go ahead and create a better-looking marker. This will especially be a good exercise when we go to create other objects later in the book.

Drag the **Map** scene back into the **Hierarchy** window and perform the following instructions to create an updated `PlaceMarker`:

1. Locate and select the `PlaceMarker` object we left in the **Map** scene and reactivate it by checking the box beside the name in the **Inspector** window.
2. Select the `Cylinder` object and rename it `Base`. In the **Inspector** window, set the objects **Transform** | **Scale** to **X**=.4, **Y**=.1, **Z**=.4, and **Transform** | **Position** to **X**=0, **Y**=-.5, **Z**=0. Remove the **Capsule Collider** component by clicking on the gear icon beside the component and selecting **Remove Component**.
3. Right-click (press *Ctrl* and click on Mac) on the `PlaceMarker`, and from the context menu, select **3D Object** | **Cylinder**. Repeat this process to create a `Sphere` and `Cube` child objects.
4. Set the properties for each of the new child objects, as shown in the following table:

Game object	Property/component	Value
Cylinder	Name	Pole
	Transform Position	(0, .5, 0)
	Transform Scale	(.05, 1, .05)
	Capsule Collider	Remove
Sphere	Name	Holder
	Transform Position	(0, 1.5, 0)
	Transform Scale	(.2, .2, .2)
	Sphere Collider	Remove
Cube	Name	Sign
	Transform Position	(0, 2, 0)
	Transform Scale	(1, 1, .1)
	Box Collider	Remove

5. Right-click (press *Ctrl* and click on Mac) on the `Assets/FoodyGo` folder in the **Project** window, and from the context menu, select **Create** | **Folder**. Rename the new folder `Materials`.
6. Select the new `Assets/FoodyGo/Material` folder in the **Project** window. Then, right-click (press *Ctrl* and click on Mac) on the empty folder, and from the context menu, select **Create** | **Material**. Rename the material `Base`. Repeat the process and create two more materials named `Highlight` and `Board`.
7. Drag the `Base` material from the **Project** window onto the `Base` object in the **Hierarchy** window. Repeat the process for the `Highlight` material onto the `Pole` and `Holder`. Finally, drag the `Board` material onto the `Sign` object.

8. The newly created materials are still the default white, so nothing visually will change. However, having the materials on the object will allow us to see the changes directly on the object as we edit the materials.
9. Select the Base material from the **Project** window, and refer to the **Inspector** window. The typical property window has changed to a shader property editor. If you are not sure what a shader is, do not worry, we will cover that in the next chapter. For now though, edit the properties in the window to match the next screenshot:

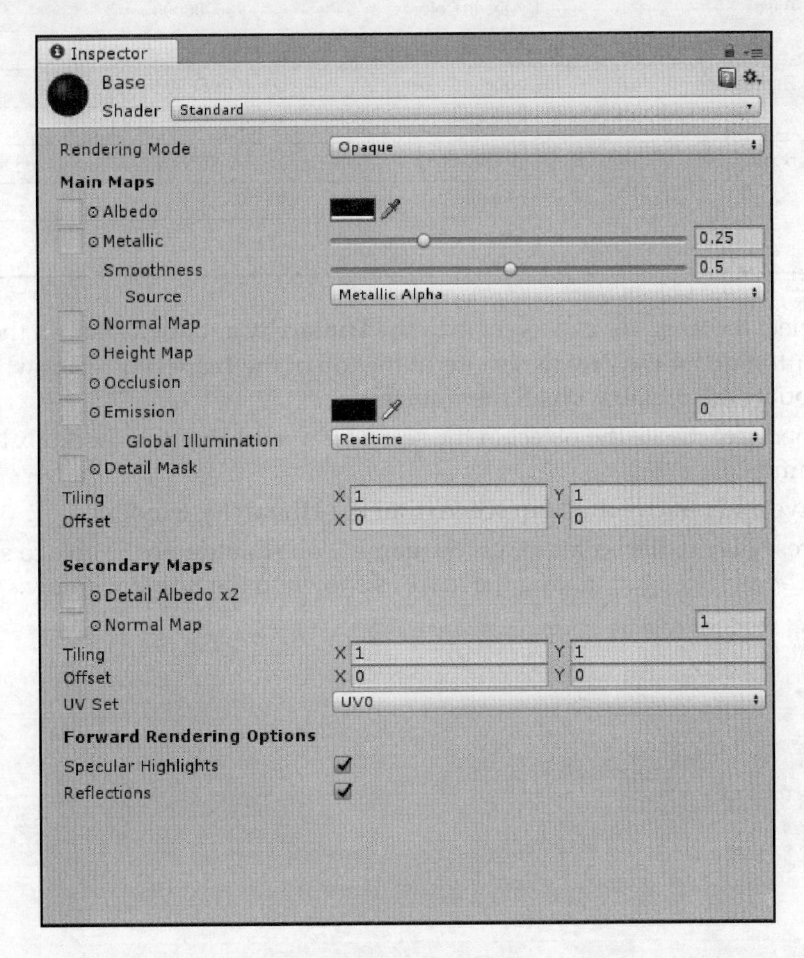

Editing the Base material shader properties

10. If you were unsure, the **Albedo** color for the `Base` material should be black (`#00000000`). Also, notice how the base of the `PlaceMarker` material properties changes as you edit the material.
11. Edit the shader material properties for each of the materials, as shown in the table:

Material	Property	Value
Base	Albedo Color	#00000000
	Metallic	.25
Highlight	Albedo Color	# 00FFE9FF
	Metallic	1
Board	Albedo Color	#090909FF
	Metallic	0
	Smoothness	0

12. Select the `PlaceMarker` prefab in the **Hierarchy** and then click on the **Apply** button under the **Prefab** options at the top of the **Inspector** window. This will update the prefab with all the changes.
13. Then, deactivate the object in the scene by unchecking the checkbox beside the name.
14. Save and remove the **Map** scene from the **Hierarchy** window.
15. Press **Play** in the editor to test the game. You should now be able to see the new `PlaceMarker` populating the map, as shown in the **Scene** window:

New PlaceMarkers shown in the Scene window

The place markers we constructed are meant to resemble a restaurant table marker and chalkboard menu combined. We will leave the menu portion blank until we allow the player to interact with the place markers, in the next chapter. For now though, we still need to resolve a couple issues with our current Google Places API search, which we will discuss in the next section.

Optimizing the search

Currently, our Google Places API search service does a nearby search and returns a page (20) of detailed results on a map redraw. Of course, if we had more than 20 locations that matched our search in an area, we would be missing several locations. We could update our requests to page through the results, but then we will start making numerous requests for every map redraw. Remember that the API we are using is not free and has tight restrictions on the number of requests.

Fortunately, there is another search option we could use to return the locations around the player. The API also supports a radar search option. A radar search will return up to 200 locations for the area we search, but the results only will contain geometry and IDs. This will work well for us, and let's edit the code to make this change and test it:

1. Double-click on the `GooglePlacesAPIService` script in the `Assets/FoodyGo/Scripts/Services` folder in the **Project** window to open the script in the editor of your choice.
2. Scroll down to the `IEnumerator SearchPlaces()` method and change the following line of code to what is here:

```
        //line to change
        var req = new WWW(GOOGLE_PLACES_NEARBY_SEARCH_URL + "?" +
queryString);

        //change to RADAR
        var req = new WWW(GOOGLE_PLACES_RADAR_SEARCH_URL + "?" +
queryString);
```

3. Save the changes to the file, and go back to the Unity editor; wait for the scripts to compile.
4. Press Play to run the game again, and notice that everything looks more or less the same as it did before.
5. Now, load the **Map** scene into the **Hierarchy** window and then navigate to the **MapScene** | **Services** | **GooglePlacesAPI** object. Change the value of the **Visual Distance** property to be 2000. Remove and save the **Map** scene.

6. Test the game in the editor again and notice that this time there are several place markers in the distance and not even on the map. Obviously, a value of 2000 is too large for our visual distance or search radius. What is the value we should be using to populate only the map with search results? The short answer is it depends.

Perhaps, if we look at our problem visually, we can determine some possible solutions to our search distance problem. Take a look at the following diagram:

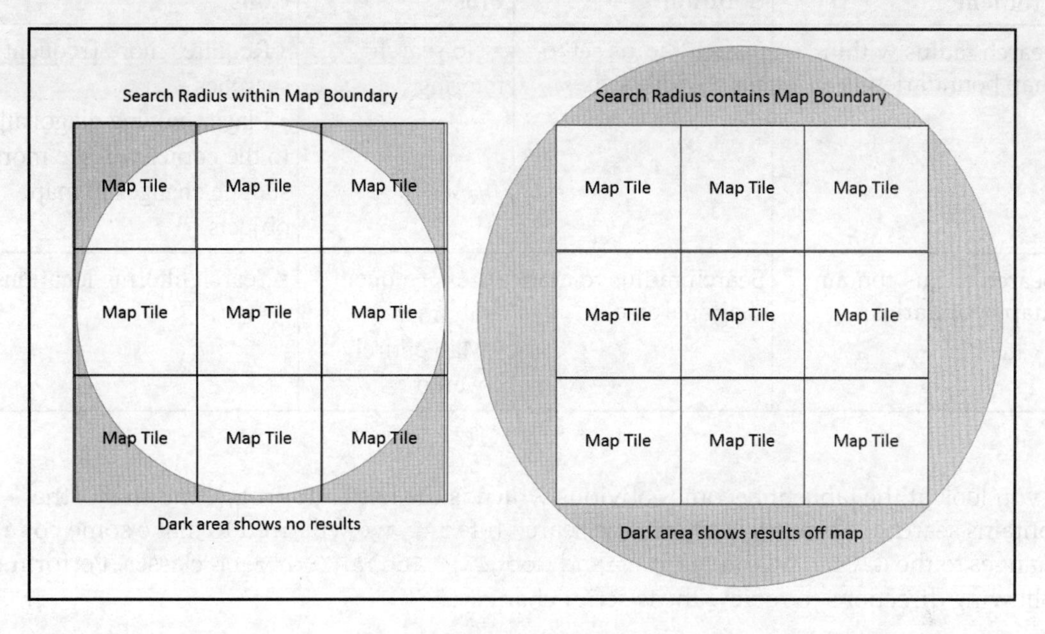

Comparison of search radius and map overlap

As the diagram shows, the problem we face is while either trying to reduce the search radius so that it is within the map boundary or it contains the map boundary. If the search radius is within our map boundary, then parts of our map will not have locations shown (the dark areas). Alternatively, if the radius contains the boundary, then the search will return a number of locations that we will be unable to map (dark areas). Ideally, we would want to be able to search by our map boundaries, but unfortunately, the Google Places API does not support this. So, we need to decide between doing a within or contains search.

Most other conventional map services provide for a map bounds query in the form of bounding box. This box is typically defined by setting the northwest and southeast points of our map as search parameters. More robust and advanced map services will also support polygon searches.

Let's breakdown the problem again in problem/solution list and then evaluate the pros and cons of each in the following table:

Problem	**Solution**	**Pros**	**Cons**
Search radius within map boundaries	Search radius set to map width size	• No search filtering	• Requires more frequent searches • Player moving diagonally to the corner will see more sudden changes in map objects
Search radius contain map boundaries	Search radius to map diagonal size	• Less frequent searching • Map entirely covered	• Search filtering locations

If you look at the table, it becomes obvious which is the better approach and that is the contains search. In order to perform the search filtering, we will need to make some code changes to the `GPSLocationService` and `GooglePlacesAPIService` classes. Perform the following directions to review those script changes:

1. From the menu, select **Assets** | **Import Package** | **Custom Package...**, and when the **Import package** dialog opens, navigate to the downloaded source code folder `Chapter_7_Assets` and select the `Chapter7_import2.unitypackage` file. We are importing the updated scripts not because there is a lot but because there are only a few changes in large files.

2. Open the `GPSLocationService` script in the editor of your choice. Then, find the `CenterMap` method. This method recalculates various important map parameters after every redraw. Added at the bottom of this file is the calculation for a new variable called `mapBounds`, as follows:

```
        lon1 = GoogleMapUtils.adjustLonByPixels(Longitude, -
MapTileSizePixels*3/2 , MapTileZoomLevel);
        lat1 = GoogleMapUtils.adjustLatByPixels(Latitude,
MapTileSizePixels*3/2 , MapTileZoomLevel);
        lon2 = GoogleMapUtils.adjustLonByPixels(Longitude,
MapTileSizePixels*3/2 , MapTileZoomLevel);
        lat2 = GoogleMapUtils.adjustLatByPixels(Latitude, -
MapTileSizePixels*3/2 , MapTileZoomLevel);
         mapBounds = new MapEnvelope(lon1, lat1, lon2, lat2);
```

3. All this code does is calculate the latitude and longitude bounds of the map. Feel free to review the section preceding this code since it has been a while (`Chapter 2`, *Mapping the Player's Location*) since we last reviewed it.
4. Next, open the `GooglePlacesAPIService` script in your editor. Find the `UpdatePlaces` method near the top of the file. The `UpdatePlaces` method is where we update the locations on the map. The method iterates through the new search results and places the objects on the map. In this method, you will find the following code:

```
        if(GPSLocationService.Instance.mapBounds.Contains(new
  MapLocation((float)lon, (float)lat))==false)
        {
            continue;
        }
```

5. That piece of code was added and uses the new `mapBounds` field on the `GPSLocationService` to determine whether the search result's latitude and longitude are within the boundaries of the map. If the search result is outside the map boundary, then the search result is ignored and the loop is continued with the continue statement.

6. Press Play to run the game. This time, move to the Scene window and zoom the camera way above the map and look down. Notice now how all the place markers are within the map boundaries, as shown in the following screenshot:

Scene view of the map showing all place markers within the map boundaries

The `GooglePlacesAPIService` now returns results that cover the entire map, and we have resolved our search problem. Ensure that you spend some time reviewing the rest of the script in order to understand how everything goes together.

Summary

In this chapter, we jumped back to the map in order to add some new real-world located features to the game. Before we added the new features, we needed to fix some of the issues that had arisen from the changes we made toward the end of the last chapter. This required us to convert our GPS service to use the Singleton pattern. As part of the conversion, we took the opportunity to understand how the Singleton worked. Then, we spent some time reviewing the Google Places API, which is the web service we would use to locate places of interest around the player. This required us to create an API key and understand how to make requests against the service with `Hurl.it`. We used `Hurl.it` to test our queries and then understood how the results, returned as JSON, could be converted into C# objects at runtime using TinyJson. With our script imported and ready, we then set up the new service within the Map scene. Then, we constructed a better prototype of our place marker using 3D primitive objects and custom materials. Finally, we determined that we needed to resolve some issues with our search, which we did with another simple script import. After that, we reviewed the changes and were happy with the final results.

With the place markers now showing on the map, we will use the next chapter to allow the player to interact with those markers by collecting objects and placing monsters. This will require us to enhance the inventory screens we developed previously. Also, we will spend some additional time enhancing the game with particle and visual effects.

8

Interacting with an AR World

Our virtual game world is now being populated with real-world-based places which we want the player to travel to and interact with. For this version of the game, we will let the player visit the places around him and try to sell his cooking monsters. If the place or restaurant likes any of the player's cooks, then it will offer a payment. The player can then accept the payment or move on. However, if the player does accept the payment, the cook will be sold to the restaurant and become its monster chef. A restaurant with a monster chef will not purchase any more cooks until the current chef leaves. A chef will leave after a certain amount of time has expired.

For this chapter, we will focus on completing that interaction of the player exploring the places around them. In order to do this, we will need to construct a new Places scene. This scene will allow the player to sell their monsters to collect new items, such as freeze balls and experience. This will require us to introduce new tables in the database, to handle new inventory items, player experience or level and track the history of a place. Also, we will be introducing a few subtle enhancements along the way. So get ready; this chapter will go over these materials very quickly and will cover the following topics:

- The Places scene
- Google Street View as a backdrop
- Slideshow with Google Places API photos
- Adding UI interaction for selling
- The game mechanics of selling
- Updating the database
- Connecting the pieces

There will be a few new elements introduced in this chapter, but most of the content added to the game at this point is an extension of concepts covered in several previous chapters. As always, we will continue the project from where we left it from the previous chapter. If you have jumped ahead from a previous chapter, consult the section on downloading the books source code for `Chapter 8`.

The Places scene

This scene will be another point where we want to mix the content of our alternate game world with the real world. Like the Catch scene, where our AR interaction was a game world on top of the real world, we will do something similar, but this time using a backdrop from Google Street View. We will also use real-world photos of the places within our markers in order to augment our game reality.

Let's get started by creating the new scene and putting the main base elements in place, as follows:

1. Open up the `FoodyGo` project from where we left in the previous chapter. If you have jumped ahead, you will need to consult the section on restoring the project from the downloaded source code.
2. Type *Ctrl* + *N* (*command* + *N* on Mac) to create a new scene.
3. Type *Ctrl* + *S* (*command* + *S* on Mac) to save the scene. Save the scene as `Places`.
4. Type *Ctrl* + *Shift* + *N* (*command* + *shift* + *N* on Mac) to create a new empty Game Object. Rename the object `PlacesScene` and reset the transform to zero.
5. Drag the `Main Camera` and `Directional Light` objects onto the `PlacesScene`, to make them child objects.
6. Select the `Main Camera` object in the **Hierarchy** window. Then, in the **Inspector** window, set the **Transform PositionY** = `2` and **Z**=`-2`. We want to raise the camera up and move it forward.
7. From the menu, select **GameObject** | **UI** | **Panel**. This will add a new `Canvas` with a child `Panel` in the **Hierarchy** window. Select the `Canvas` object.
8. Locate the **Canvas** component in the **Inspector** window, and set the **Render Mode** to **Screen Space – Camera**. Then, drag the `Main Camera` object from the **Hierarchy** window into the empty **Render Camera** slot on the **Canvas** component.
9. Select the `Panel` object in the **Hierarchy** window. Then, in the **Inspector** window, change the **Image** component **Color** to `#FFFFFFFF` by clicking on the color slot to open the **Color** dialog and enter the hex value.

10. Locate the `PlaceMarker` prefab in the `Assets/FoodyGo/Prefabs` folder in the **Project** window. Select the prefab and type *Ctrl* + *D* (*command* + *D* on Mac) to duplicate the prefab. Rename the prefab `PlacesMarker` and drag the new prefab into the **Hierarchy** window.
11. Finally, drag the `Canvas` and `PlacesMarker` objects onto the `PlacesScene` object in the **Hierarchy** window to make them children. Your **Game** and **Hierarchy** windows should match the following screenshot:

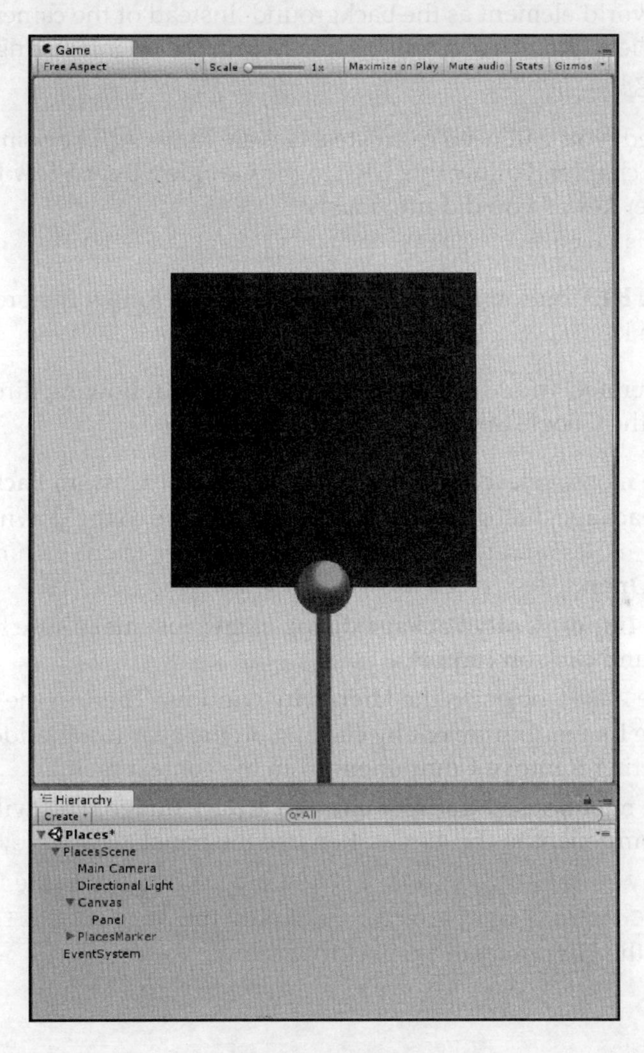

The start of the Places scene

That sets up our base **Places** scene, and now, we want to add the other AR elements, such as the backdrop, and sign in the next sections.

Google Street View as a backdrop

Much like the Catch scene we created earlier, which used the camera as a backdrop, we want to use a real-world element as the background. Instead of the camera, this scene will use Google Street View. Street View will be able to provide an interesting backdrop for the scene without using the camera.

Before we get started, you will need to create a Google Maps API key similar to what you did in the previous chapter. Follow this URL to the **Google Street View Image API** to generate a developer key as you did previously:
`https://developers.google.com/maps/documentation/streetview/`.

Click on the **GET A KEY** button at the top of the page, and follow the process as you did before to generate a key.

After you have generated your developer key, perform the following directions to set up the backdrop with the Google Street View Image API:

1. From the menu, select **Assets** | **Import Package** | **Custom Package...** to open the **Import package** dialog. From the dialog, navigate to the downloaded source code `Chapter_8_Assets` folder. Select the `Chapter8_import1.unitypackage` and click on **Open**.
2. After the **Import Unity Package** dialog opens, just make sure that everything is selected and click on **Import**.
3. Select the `Panel` object in the **Hierarchy** window. Then, in the **Inspector** window, delete the Image component by clicking on the gear icon beside the component and selecting **Remove Component** from the context menu.
4. From the menu, select **Component** | **UI** | **Raw Image**. This will add a new **Raw Image** component to the `Panel`. Rename the panel `StreeViewTexturePanel`.
5. From the `Assets/FoodyGo/Scripts/Mapping` folder in the **Project** window, drag the `GoogleStreetViewTexture` onto the `StreetViewTexturePanel` object in the **Hierarchy** or **Inspector** window.

6. In the **Inspector** window, set the properties of the **Google Street View Texture** component to the values, as follows:

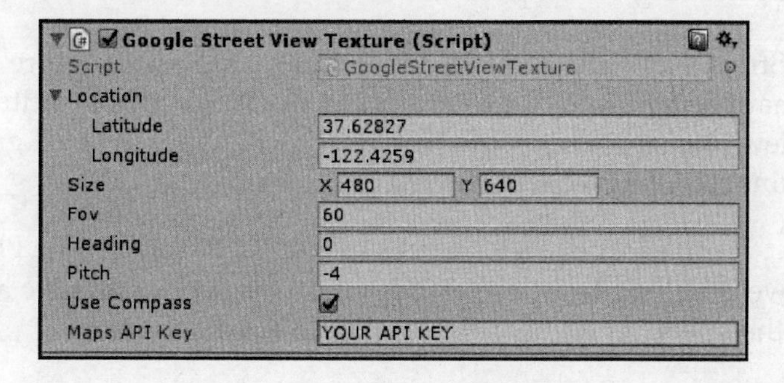

Google Street View Texture component configuration

7. Press Play to run the game in the editor and notice how the background is now the street outside the Google office.
8. From the menu, select **File** | **Build Settings...**, and click on the **Add Open Scenes** button to add the `Places` scene to **Scenes in Build**. Then, uncheck the other scenes, as shown in the **Build Settings** dialog:

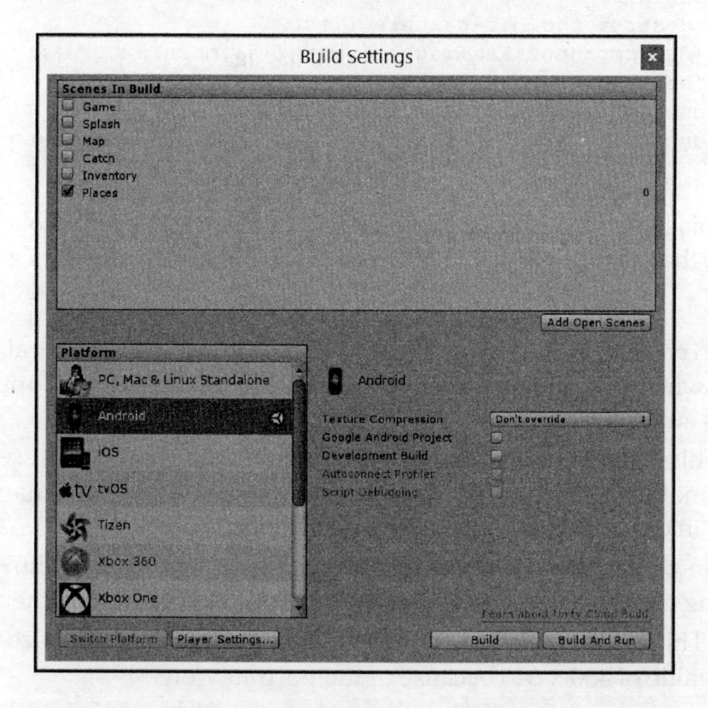

Updated build settings for running the Catch scene on a device

9. Build and deploy the game to your mobile device. As the scene is running, rotate the device and watch as the background changes. The background changes to match the heading on the phone.

Using the Google Street View Image API is great as a background texture when the player engages in areas mapped by Street View. However, not all locations (latitude/longitude) will have Street View imagery. This could be fixed by changing the background for locations with no imagery, but we will leave it as is for now.

The `GoogleStreetViewTexture` script works much like to the Google Places and Google Maps API scripts we developed earlier. Let's open up the script and take a look at the main method that does the query, as follows:

```
IEnumerator LoadTexture()
{
    var queryString =
string.Format("location={0}&fov={1}&heading={2}&key={3}&size={4}x{5}&pitch=
{6}", location.LatLong, fov, heading, MapsAPIKey, size.x, size.y, pitch);

    var req = new WWW(GOOGLE_STREET_VIEW_URL + "?" + queryString);
    //yield until the service responds
    yield return req;
    //first destroy the old texture first
    Destroy(GetComponent<RawImage>().material.mainTexture);
    //when the image returns set it as the tile texture
    GetComponent<RawImage>().texture = req.texture;
    GetComponent<RawImage>().material.mainTexture = req.texture;
}
```

As mentioned, this looks very similar to the other code we used earlier to access the other Google APIs. So, the main thing we will focus on is the query parameters getting submitted to the API:

- `location`: This is the latitude and longitude coordinates separated by a comma. Later, when we connect this script to the GPS service, it will populate these values automatically.
- `fov`: This is the field of view, in degrees, of the image, which essentially equates to zoom. We use a value of 60, which represents a reasonable narrow view of our world in order to match the device's screen dimensions.
- `heading`: This is the compass heading in degrees north. We allow the compass heading of the device set this value if the option is enabled, otherwise 0.
- `size`: This is the size of the requested image. The maximum size is 640 x 640. We use a value of 430 x 640 because of the portrait view.

- `pitch`: This is the angle above or below the horizon. Values can range from -90 to 90. We use a value -4 to get a bit of the ground in the Street View images.
- `key`: This is the API key you generated for the Google Street View API.

Feel free to review the rest of the `GoogleStreetViewTexture` script on your own. You will find the rest of the code similar to what we did in the Google Places and Maps scripts. In the next section, we will cover additional queries against the Google Places API.

Slideshow with the Google Places API photos

In the previous chapter, we used the Google Places API in order to populate the place markers around the map with a radar search. Now, as we have the player interacting with a marker, we want them to be able to visualize information about the location. We can do that by executing a detail query about the place, which will return additional information as well as the attached photos. Those photos will be used to create a slideshow for the player while visiting the place.

Along with the slideshow, we will also add text about the place name and the current rating. Later, we will use the rating to determine what cooking monsters a place will want to purchase and for how much.

You will need the Google Places API key that we generated in the previous chapter in order to complete the configuration in the next section. If you jumped ahead from an earlier chapter, review the process for generating the API key in the previous chapter.

We will start by adding the new elements we need for the scene by following the instructions here:

1. Return to the **Places** scene and locate the `PlacesMarker` in the **Hierarchy** window.
2. Expand the `PlacesMarker` and select the `Sign` object in the **Hierarchy** window. Press *Ctrl* + *D* (*command* + *D* on Mac) to duplicate the object. Rename the new object **Photo**.

3. With the new `Photo` object selected in the **Inspector** window, change the **Mesh Renderer** | **Material** by clicking on the bullseye icon to open the **Select Material** dialog. Then, scroll to the bottom of the list and select **Default-Material**. After making the selection, close the dialog and confirm that the material has been replaced.
4. Change the **Transform** properties of the `Photo` object to match the following screenshot:

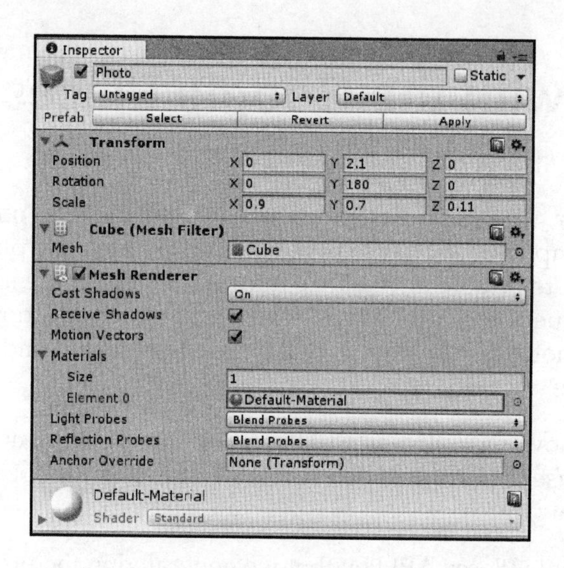

Photo object properties

5. From the menu, select **GameObject** | **UI** | **Canvas** to create a new `Canvas` object. Drag the `Canvas` object onto the `PlacesMarker` object in the **Hierarchy** window. We added the canvas to the object so that we can add text directly onto the object.

6. With the new `Canvas` selected in the **Inspector** window, change the **Canvas** | **Render Mode** to **World Space** and then change the other properties to match the values, as shown in the following screenshot:

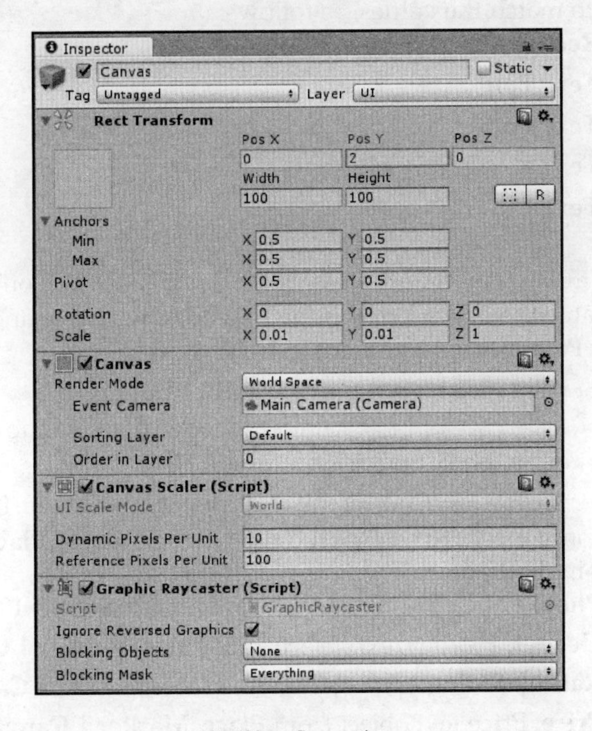

World Space Canvas settings

You will need to manually set all the properties on the **Rect Transform** after you change **Canvas** | **Render Mode** to **World Space**.

7. Right-click (press *Ctrl* and click on Mac) on the `Canvas` object, and from the context menu, select **UI** | **Text**. Rename the new element **Header**.
8. In the **Inspector** window, change the properties of the `Heading` object to match the values:
    - **Rect Transform**: **Pos Z** = `-.1`
    - **Rect Transform**: **Width** = `90`, **Height** = `70`
    - **Text**: **Font Size** = 10
    - **Text**: **Paragraph Alignment** = `Left and Bottom`
    - **Text**: **Color** = `#FFFFFFFF (white)`

9. Select the `Header` object in the **Hierarchy** window and press *Ctrl* + *D* (*command* + *D* on Mac) to duplicate the object. Rename the object **Rating.**
10. Select the new `Rating` object, and in the **Inspector** window, change the properties to match the values, as follows:
    - **Rect Transform: Height** = `90`
    - **Text: Font Size** = `7`
    - **Text: Font** = `fontawsome-webfont`
    - **Text: Paragraph Alignment** = `Right and Bottom`
    - **Text: Color** = `#00FFFFFF (cyan)`
11. Select the `Rating` object and press *Ctrl* + *D* (*command* + *D* on Mac) to duplicate the object. Rename the object **Price** in the **Inspector** window, and set the **Text** component **Paragraph Alignment** to **Left – Bottom.**
12. Drag the `GooglePlacesDetailInfo` script from the `Assets/FoodyGo/Scripts/UI` folder in the **Project** window onto the `PlacesMarker` in the **Hierarchy** window.
13. Select the `PlacesMarker` object in the **Hierarchy** window. Then, drag the following components to each of the slots in the **Google Places Detail Info** component in the **Inspector** window, as follows:
    - **Photo Panel**: Photo object from **PlacesMarker** | **Photo**
    - **Header**: Header text object from **PlacesMarker** | **Canvas** | **Header**
    - **Rating**: Rating text object from **PlacesMarker** | **Canvas** | **Rating**
    - **Price:** Price text object from **PlacesMarker** | **Canvas** | **Price**
14. Set the other properties on the component to the values, as shown in the following screenshot:

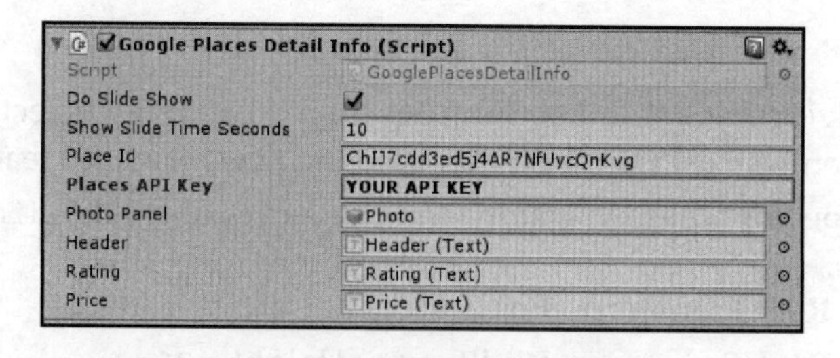

Google Places Detail Info component settings

The **Place Id** value used in this example is (`ChIJ7cdd3ed5j4AR7NfUycQnKvg`). You could also use other place ID values by doing a manual Google Places API Nearby search, like we did in the previous chapter.

15. Finally, save the `PlacesMarker` prefab by clicking on the **Apply** button beside the **Prefab** options in the **Inspector** window.
16. Press Play to run the scene. The scene is somewhat fabricated, but it will show you the functionality we have added. Wait for a while and watch as the photos on the sign change in a slideshow. The following is a sample screenshot from the scene running in the **Game** window:

The Places scene running in Game window

As you can see, our scene is progressing nicely. Let's quickly take a look at a couple of main methods in the `GooglePlacesDetailInfo` script. We will start with the `LoadPlacesDetail` method:

```
IEnumerator LoadPlacesDetail()
{
    var queryString = string.Format("placeid={0}&key={1}"
                , placeId, PlacesAPIKey);

    var req = new WWW(GOOGLE_PLACES_DETAIL_URL + "?" + queryString);
    //yield until the service responds
    yield return req;
    var json = req.text;
     ParseSearchResult(json);
}
```

By now, the code within these `Coroutine` methods should be old hat. This code is very similar to what we used in order to make the Google Places Radar search in the previous chapter. As you can see in the first line, where the `querystring` variable is getting created, the URL parameters are `placeid` and key, where `placeid` is the `id` of the location and key is, of course, your Google Places API key.

After the script does a request for the detail information about a place, it sets the header text and rating. If the place has any photos, it runs a new query against the Google Places API Photos endpoint, as shown in the `LoadPhotoTexture` method:

```
private IEnumerator LoadPhotoTexture(Photo photo)
{
    var queryString =
string.Format("photoreference={0}&key={1}&maxwidth=800"
                , photo.photo_reference, PlacesAPIKey);

    var url = GOOGLE_DETAIL_PHOTO_URL + "?" + queryString;
    var req = new WWW(url);
    //yield until the service responds
    yield return req;
    //first destroy the old texture first
    Destroy(photoRenderer.material.mainTexture);
    //when the image returns set it as the tile texture
    photoRenderer.material.mainTexture = req.texture;
}
```

This code is very close to the loading of the Street View textures we did earlier in the chapter. The only main difference is the parameters we pass. This endpoint only requires three parameters: a `photoreference` (essentially an ID), `key` (that API key again), and a `maxwidth` (or `maxheight`). The `photoreference` value is extracted from the main query that was executed previously.

The last method we will look at is the `Coroutine` that runs the slideshow, called `SlideShow`:

```
private IEnumerator SlideShow(Result result)
{
    while(doSlideShow && idx < result.photos.Count - 1)
    {
        yield return new WaitForSeconds(showSlideTimeSeconds);
        idx++;
        StartCoroutine(LoadPhotoTexture(result.photos[idx]));
    }
}
```

The `Result` parameter passed to this method is the object result, which describes the place. Inside this method is a continuous while loop that checks whether the `doSlideShow` is true and whether the `result` has any photos left to show. `idx` is an index to the current photo being shown, starting at 0. As you can see, the result variable has an `Array` property called `photos`, which holds all the `photoreference` values of photos taken for that place. In the `while` loop, we compare the current index (`idx`) with the number of photos. If there are more photos, the loop executes. The first statement of the loop causes the routine to yield for a set period defined by `showSlideTimeSeconds`. After that, it increments the `idx` and then loads another photo.

With our backdrop and place sign working, we can now add the monster-selling interaction in the next section.

Adding UI interaction for selling

As described at the beginning of the chapter, players will sell their monsters by going to a place and then initiating a sale by clicking on a button. The place or restaurant will then examine the player's monsters and offer a price for what it considers the best. A player can then accept the offer by clicking on a **Yes** button or refuse by clicking on a **No** button. Even with this very simple workflow, there are still a lot of elements we will need to put together, most of which will be UIs.

We will break down building the UI into sections so that it is easier to follow. The first part we will work on is adding the UI canvas and the primary buttons by performing the following directions:

1. From the menu, select **GameObject | UI | Canvas**. Rename the new canvas **UI_Places**. If you have been counting, that is our third canvas in the scene, with each rendering to a different space (Screen Space-Overlay, Screen Space-Camera and World Space).
2. Select the new canvas, and in the **Inspector** window, change the **Canvas Scaler** component properties, as follows:
    - **Canvas Scaler**: **UI Scale Mode** = `Scale with Screen Size`
    - **Canvas Scaler**: **Reference Resolution**: **X** = `500`, **Y** = `900`
3. Right-click (press *Ctrl* and click on a Mac) on the `UI_Places` object, and from the context menu, select **UI | Button**. Rename the new button `SellButton`.
4. Expand the `SellButton` in the **Hierarchy** window, and select the `Text` object. Type *delete* to remove the `Text` object from the button.
5. Select the `SellButton` in the **Hierarchy** window and type *Ctrl* + *D* (*command* + *D* on a Mac) to duplicate the button. Rename the new button **ExitButton**.
6. For each of the new buttons, change the component properties according to the following table:

Button	Property	Value
SellButton	Rect Transform - Anchors	Top-Center pivot and position
	Rect Transform – Pos Y	-130
	Rect Transform – Width, Height	100
	Image – Source Image	button_set03_a
ExitButton	Rect Transform - Anchors	Bottom-Center pivot and position
	Rect Transform – Pos Y	10
	Rect Transform – Width, Height	75
	Image – Source Image	Button_set11_b

That completes adding the buttons to interface. Don't worry about wiring up the buttons, yet. We will instead work to complete the rest of the UI elements needed in the scene. Next, we will construct the `OfferDialog`:

1. Right-click (press *Ctrl* and click, on a Mac) on the `UI_Places` canvas in the **Hierarchy** window, and from the context menu, select **UI** | **Panel**. Rename the panel **OfferDialog**. This will be the parent object for all our dialog components.
2. Select the `OfferDialog` panel in the **Hierarchy** window, and then from the menu, select **Component** | **Layout** | **Vertical Layout Group**. Then, from the menu, select **Component** | **UI** | **Effects** | **Shadow**.
3. In the **Inspector** window, set the **Rect Transform**, **Image**, and **Vertical Layout Group** component properties, as shown in the next screenshot:

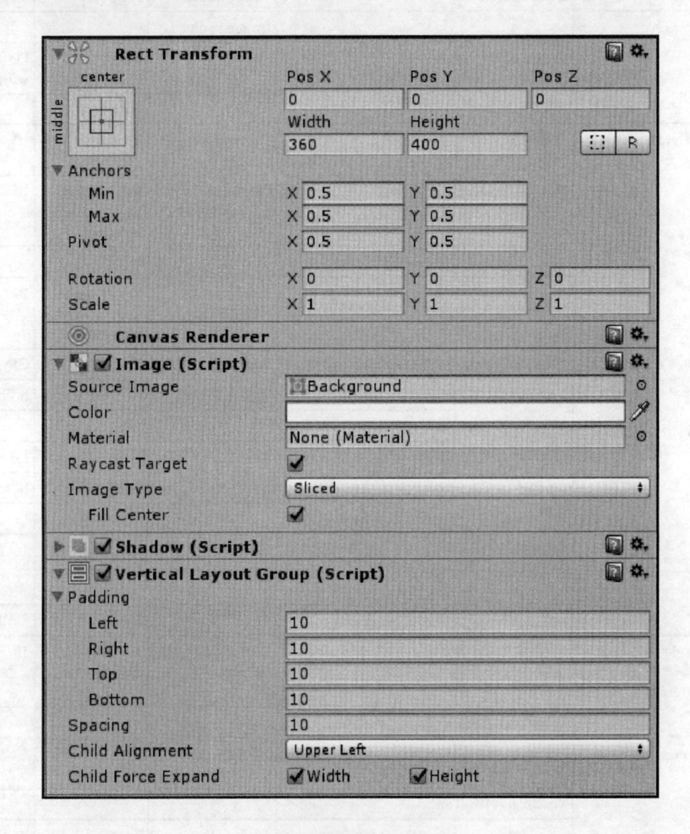

OfferDialog component properties

4. Right-click (press *Ctrl* and click on a Mac) on the `OfferDialog` in the **Hierarchy** window, and from the context menu, select **UI | Panel**. Rename the panel `PromptPanel`.
5. Select the `PromptPanel` in the **Hierarchy** window and press *Ctrl* + *D* (*command* + *D* on a Mac) to duplicate the object. Do this two more times so that there are three new panels in total. Rename the new panels as follows: `MonsterDetailPanel`, `OfferPanel`, and `ButtonPanel`.
6. Notice how the panels inside the layout group adjust to evenly fit the space. Now that you have an idea of how to add layouts and build a UI hierarchy, continue building the dialog, by following the table:

Panel	UI Hierarchy		Component	Property	Value
PromptPanel			Image : remove		
	PromptText [Text]		Text	Text	Do you want to sell?
			Text	Font Size	18
MonsterDetailPanel			Vertical Layout Group	Spacing	10
	HeaderPanel [Panel]		Horizontal Layout Group	Left, Right, Top, Bottom, Spacing	10
		NameText [Text]			
	DescriptionPanel [Panel]		Horizontal Layout Group	Left, Right, Top, Bottom, Spacing	10
		CP [Text]	Text	Text	CP:
			Text	Font Style	Bold
		CPText [Text]			
		Level [Text]	Text	Text	Level:
			Text	Font Style	Bold
		LevelText [Text]			
	SkillsPanel [Panel]		Horizontal Layout Group	Left, Right, Top, Bottom, Spacing	10
		SkillsText [Text]	Text	Text	Skills
OfferPanel			Vertical Layout Group	Left, Right, Top, Bottom, Spacing	10
	OfferText [Text]				
ButtonPanel			Horizontal Layout Group		
	YesButton [Button]				
		Text	Text	Text	Yes
	NoButton [Button]				
		Text	Text	Text	No

7. Check whether your final construction matches the dialog and hierarchy:

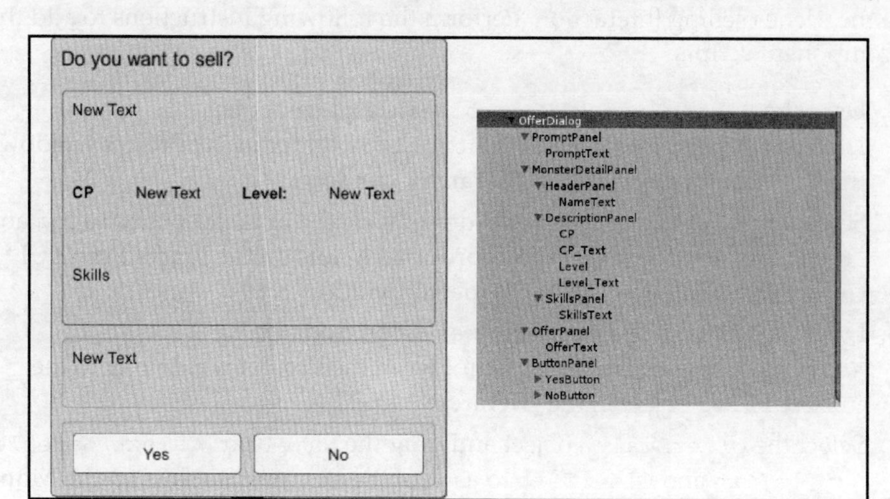

Completed OfferDialog and object hierarchy

With the `OfferDialog` completed, we can quickly add the `RefuseDialog` by following the next directions:

1. Select the `OfferDialog` panel in the **Hierarchy** window and press *Ctrl* + *D* (*command* + *D* on Mac) to duplicate the dialog. Rename the panel `RefuseDialog`.
2. In the **Inspector** window, change the **Rect Transform – Height** = `150`.
3. Expand the `RefuseDialog`, and delete the `MonsterDetailPanel` and `OfferPanel`.
4. Expand the `PromptPanel` and then select the `PromptText` in the **Hierarchy** window. In the **Inspector** window, change the **Text – Text** to `This place will not be making an offer for any of your current stock!`.
5. Expand the `ButtonPanel` in the **Hierarchy** window, and delete one of the buttons, it doesn't matter which. Rename the remaining button as `OKButton`. Expand the button, and select the child `Text` object. In the **Inspector** window, change the **Text – Text** to `OK`.
6. Finally, drag the `UI_Places` object onto the `PlacesScene` object in the **Hierarchy** window to make it a child of the scene root.

With the UI elements all constructed, we can now wire up all the scripts we need to complete the scenes selling interaction. Perform the following instructions to add the various component scripts:

1. Drag the `PlacesSceneUIController` script from the `Assets/FoodyGo/Scripts/Controllers` folder in the **Project** window onto the `UI_Places` object in the **Hierarchy** window.
2. Select the `UI_Places` object and drag the `SellButton`, `OfferDialog` and `RefuseDialog` to each of the appropriatee slots on the **Places Scene UI Controller** component in the **Inspector** window.
3. Drag the `MonsterOfferPresenter` script from the `Assets/FoodyGo/Scripts/UI` folder in the **Project** window onto the `OfferDialog` object in the **Hierarchy** window.
4. Select the `OfferDialog` object and drag the `NameText`, `CPText`, `LevelText`, `SkillsText` and `OfferText` to each of the appropriate slots on the **Monster Offer Presenter** component in the **Inspector** window.
5. Drag the `PlacesSceneController` script from the `Assets/FoodyGo/Scripts/Controllers` folder in the **Project** window onto the `PlacesScene` root object.
6. Select the `PlacesScene` object and drag the `StreeViewTexturePanel`, `PlacesMarker`, and `UI_Places` objects to the appropriate places on the **Places Scene Controller** component in the **Inspector** window, as shown in the following screenshot:

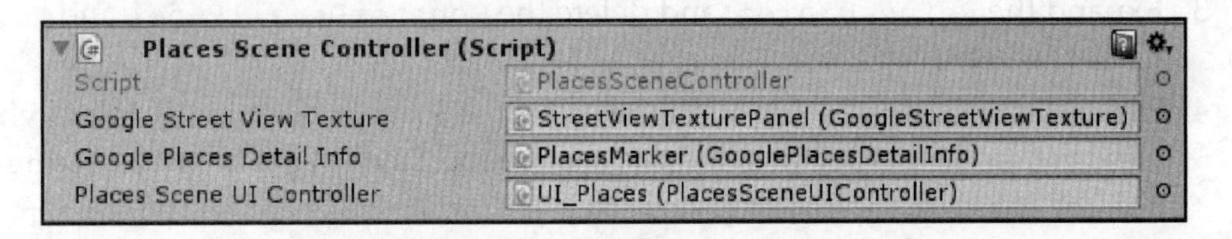

The Places Scene Controller component object settings

7. Select the `ExitButton` in the **Hierarchy** window, and then in the **Inspector** window, add a new event handler by pressing the plus. Then, drag the `PlacesScene` object onto the object slot and then open the function dropdown. Select the **PlacesSceneController.OnCloseScene** function, as follows:

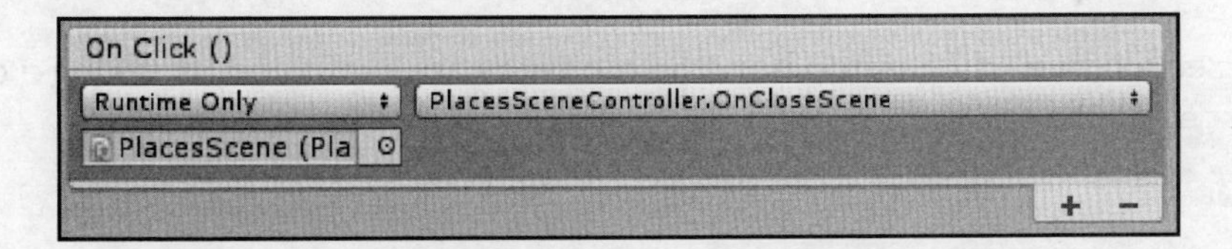

The ExitButton event handler configuration

8. Repeat the process for the `SellButton`, but this time select the **PlacesSceneController.OnClickSell** function from the function dropdown.
9. Select the `YesButton` and add a new event handler. Drag the `UI_Places` object into the object slot, and select the `PlacesSceneUIController.AcceptOffer()` for the function.
10. Select the `NoButton` and add a new event handler. Drag the `UI_Places` object into the object slot, and select the `PlacesSceneUIController.RefuseOffer()` for the function.
11. Select the `OKButton` and add a new event handler. Drag the `UI_Places` object into the object slot, and select the `PlacesSceneUIController.OK()` for the function.

That finishes wiring up the scene. Now, we just need to add a couple of services in order to run the Places scene by itself. Perform the following directions to add the services:

1. Create a new empty game object called `Services` in the scene. Drag it onto the `PlacesScene` object in order to make it a child of the root scene object.
2. Add two new child objects to the `Services` objects called `Inventory` and `MonsterExchange`.
3. Drag the `InventoryService` script onto the `Inventory` object and the `MonsterExchangeService` script onto the `MonsterExchange` object from the `Assets/FoodyGo/Scripts/Services` folder in the **Project** window.
4. Press Play to run the scene, and try clicking on the Sell button and various other buttons. As you will notice, you can't actually sell a monster yet, but the UI should all be working as expected, except for the `ExitButton`, of course.

That completes all the required UI elements for the scene, which as you have seen is quite a lot for a simple scene. The UI is still very basic, and it will be left up to you, the developer, to extend and make your own later. Feel free to explore and change the UI dialogs the way you like. You may also notice that we do not provide feedback to the player if they do sell items. That is left up to the developer to incorporate a new dialog or sound as they see fit.

In the next section, we will get into how the monsters are evaluated by taking a closer look at the `MonsterExchangeService`.

The game mechanics of selling

Of the many things game developers can take pleasure in, most certainly creating game mechanics and game algorithms has to be at the top of the list. There is no clearer way to distinguish your game or virtual world than from the game mechanics that drive it. As a game developer and designer, you have many options for defining game mechanics, be it creating powerful weapons or creatures, world physics, or even just selling cooking monsters. Generally though, it is best to keep things fairly simple, and we have stuck with this philosophy when building the selling game mechanic.

Instead of looking at the actual source code of the `MonsterExchangeService`, which is the service that evaluates the player's monster inventory. We will look at some more basic pseudo code that will be easier to follow:

```
MonsterOffer PriceMonsters(PlaceResult result)
{
    //get the players monsters from Inventory
    var monsters = InventoryService.ReadMonsters();
    //loop through all the reviews of the place
    //and append into a single text string
    var reviews = string.empty;
    foreach(var r in result.reviews)
    {
        reviews += r.text + " ";
    }

    //loop through the monsters and place a value to each
    List<MonsterOffer> offers = new List<MonsterOffer>();

    //calculate the places budget to spend on cooking monsters
    var budget = result.rating * result.price * result.types * 100;

    foreach(var m in monsters)
    {
        var value = 0;
```

```
        //add a bonus value based on every skill word
       //that matches a word in the review
        //ex. skill=pizza, reviews may mention the word pizza 5 times = 500
        value += CountMatches(m.skills, reviews)*100;
        //add the cooking power (CP)
        value += m.power * m.level * 100;
        if(budget > value)
            //add a new offer for the monster
            offers.Add(new MonsterOffer{ Monster = m, Value = value });
    }
    offers.sort();  //sort the offers from lowest to highest
    offers.reverse();  //reverse the order so highest on top

    return offer[0];   //return the top offer
}
```

The algorithm we are using here is fairly simple with one twist where we actually try to give a bonus to monsters that may have skills mentioned in the place reviews. If you open up the `MonsterExchangeService` script, you will notice that the code follows the same pattern. Obviously, you are free to modify this gam mechanic in any manner you like.

e mechanic in any manner you like. After we create a numeric value for the cooking monster, we don't worry about converting that number into game terms until the offer is presented to the player. When the offer is presented to the player, the number then gets converted to experience and items by the `MonsterExchangeService` using the `ConvertOffer` method.

We won't get into the details of the `ConvertOffer` method here, as the code just breaks the value up by boundaries similar to how a coin change algorithm may work. The following is a list showing the boundaries as they are defined in the script:

- 1000: Nitro-Ball
- 250: DryIce-Ball
- 100: Ice-Ball

As an example, if an offer was valued at 2600, then the `ConvertOffer` would break down that value into:

- 2: Nitro-Balls (2000)
- 2: DryIce-Balls (500)
- 1: Ice-Ball (100)
- 260 Experience: experience is calculated at 1/10 the value

Again, feel free to open up the MonsterExchangeService script and add new items or change the values. Now that we understand how the game mechanics of how a cooking monster is valued, we need to move on to actually making the exchange, which is covered in the next section.

Updating the database

As you likely noticed when testing the Places scene UI, you can't actually sell a monster just yet. That is because we need to add a few new tables to our database (Inventory). Fortunately, as you may recall that since our database is an ORM (object relational mapping), the work of creating new tables is quite painless.

So, open up the InventoryService script located in the Assets/FoodyGo/Scripts/Services folder, in the editor of your choice. Scroll down to the CreateDB method, and look for the new section following code after the highlighted line to create the new tables in the database:

```
Debug.Log("DatabaseVersion table created."); //look for this line to start
 //create the InventoryItem table
 var iinfo = _connection.GetTableInfo("InventoryItem");
 if (iinfo.Count > 0) _connection.DropTable<InventoryItem>();
 _connection.CreateTable<InventoryItem>();
 //create the Player table
 var pinfo = _connection.GetTableInfo("Player");
 if (pinfo.Count > 0) _connection.DropTable<Player>();
 _connection.CreateTable<Player>();
```

That code just adds the new tables, InventoryItem, and Player to the database. Next, scroll down a little further in the same method, and look at the code required to create a new starting player after the highlighted line:

```
Debug.Log("Database version updated to " + DatabaseVersion);
//start here
_connection.Insert(new Player
{
   Experience = 0,
   Level =1
});
```

After adding the new tables in the `CreateDB` and populating with a starting `Player`, the `UpgradeDB` method was updated as well. The old `UpgadeDB` was replaced with the following code:

```
private void UpgradeDB()
{
    var monsters = _connection.Table<Monster>().ToList();
    var player = _connection.Table<Player>().ToList();
    var items = _connection.Table<InventoryItem>().ToList();
    CreateDB();
    Debug.Log("Replacing data.");
    _connection.InsertAll(monsters);
    _connection.InsertAll(items);
    _connection.UpdateAll(player);
    Debug.Log("Upgrade successful!");
}
```

If a database upgrade is triggered, then the `UpgradeDB` method will run. It will first store the values in the current tables in temporary variables, then the `CreateDB` method is run to create the new database tables and finally that data is inserted or updated back in the database. Being able to upgrade the database in this manner allows us to add new properties to an object. However, we can never delete or rename existing properties as that would likely break the old objects.

Now that the tables are in place, we can add the CRUD methods for our new tables, `InventoryItem` and `Player`. Scroll to the bottom of the `InventoryService` file and look at the following new methods:

```
//CRUD for InventoryItem
public InventoryItem CreateInventoryItem(InventoryItem ii)
{
     var id = _connection.Insert(ii);
     ii.Id = id;
     return ii;
}

    public InventoryItem ReadInventoryItem(int id)
{
    return _connection.Table<InventoryItem>()
                .Where(w => w.Id == id).FirstOrDefault();
}

public IEnumerable<InventoryItem> ReadInventoryItems()
{
    return _connection.Table<InventoryItem>();
}
```

```
public int UpdateInventoryItem(InventoryItem ii)
{
    return _connection.Update(ii);
}

public int DeleteInventoryItem(InventoryItem ii)
{
    return _connection.Delete(ii);
}
```

The code is almost the same as the Monster CRUD code we wrote in a previous chapter. Now, look at the CRUD code for the `Player`:

```
//CRUD for Player
public Player CreatePlayer(Player p)
{
    var id = _connection.Insert(p);
    p.Id = id;
    return p;
}

public Player ReadPlayer(int id)
{
   return _connection.Table<Player>()
                .Where(w => w.Id == id).FirstOrDefault();
}

public IEnumerable<Player> ReadPlayers()
{
    return _connection.Table<Player>();
}

public int UpdatePlayer(Player p)
{
    return _connection.Update(p);
}

public int DeletePlayer(Player p)
{
    return _connection.Delete(p);
}
```

You may be wondering why we need all the CRUD methods for the **Player**. We put them there so that our game could be used to handle multiple players later either by allowing a user to choose different characters to play with or perhaps even allowing multiple players to play together.

Next, we want to jump back to the **Places** scene and run an upgrade to our database so that the new tables are all there. Ensure that you save all your scripts, and go back to the Unity editor. Find the `InventoryService` object and select it in the **Hierarchy** window. In the **Inspector** window, change the **Inventory Service – Database Version** to **1.0.1** (assuming the current value is 1.0.0). Run the **Places** scene in the editor, and check the **Console** window. Look for the **Upgrade Successful** log message to ensure that your database is upgrading correctly.

Anytime you want to upgrade the database in the future, just increment the version to a higher value. Version 1.0.3 is higher than 1.0.1, and version 2.0.0 is higher than 1.0.22.

With the `InventoryService` code and database updated, now we need to perform the exchange of the player where they will give the monster to the place and receive their experience and items in return. Open up the `PlacesSceneUIController` script in your editor and scroll down to the `AcceptOffer` method:

```
public void AcceptOffer()
{
    OfferDialog.SetActive(false);
    SellButton.SetActive(true);

    var offer = CurrentOffer;
    InventoryService.Instance.DeleteMonster(offer.Monster);
    var player = InventoryService.Instance.ReadPlayer(1);
    player.Experience += offer.Experience;
    InventoryService.Instance.UpdatePlayer(player);
    foreach(var i in offer.Items)
    {
        InventoryService.Instance.CreateInventoryItem(i);
    }
}
```

The top two lines of this method just enable/disable dialog and button. Then, we use those CRUD methods to remove the monster from the inventory. Next, we update the player's experience and then add the items to the new `InventoryItem` table. We won't be using those new items in this chapter, but we now have a place to store new inventory.

Return back to the Unity, and run the **Places** scene. Now, try selling the monsters. You should notice that monsters are getting sold to the place. Verify this by going back and continually selling more monsters. Each time you sell a new monster, they will likely have a new name and other skills.

While you are running the game from the editor, the Inventory Service will make sure that there is always a random monster in the Inventory. This is done for debugging purposes. However, if you do encounter that a place is not able to buy your last monster, this can stall your testing. Consult `Chapter 10`, *Troubleshooting*, for tips on how to inspect and modify the database directly.

Our **Places** scene is complete for now. Feel free to extend the elements in the scene to however you desire. In the next section, we will add the **Places** scene back into the game and allow the player to visit a place.

Connecting the pieces

With the Places scene complete, now we need to integrate the scene back into the Gamescene. Perform the following instructions to configure the Places scene so that it can be loaded by the `GameManager`:

1. From the `Assets` folder in the **Project** window, drag the `Game` scene into the **Hierarchy** window.
2. Select the `Inventory` service in the **Places** scene, and note what you set the **Inventory Service – Database Version** to in the **Inspector** window. Then, select the `Inventory` service in the Game scene and change the **Inventory Service – Database Version** to match.

If we did not do this step, the database may get corrupted and just wipe itself clean with the updated objects. This is not so critical at this point, but would be very inconvenient to any users if you were creating an update of the game.

3. Since we only need one `Inventory` service, select the `Inventory` service object in the **Places** scene and press *delete* to remove it. Select and delete the `Event System` object in the **Places** scene also.
4. Select the `PlacesScene` object in the **Hierarchy** window and deactivate it by unchecking the checkbox beside the object name in the **Inspector** window. We will disable the root object because we will be loading the scene on start.
5. Right-click (press *Ctrl* and click on a Mac) on the **Places** scene, and select **Remove Scene** from the context menu. Ensure that you save the changes we just made to the scene.
6. Select the `_GameManager` object in the **Hierarchy** window. Set the **Game Manager – Places Scene Name** property to `Places` in the **Inspector** window.
7. From the menu, select **File | Build Settings** to open the **Build Settings** dialog. Just confirm that all the scenes have been added and are checked to be included, as shown in the following screenshot:

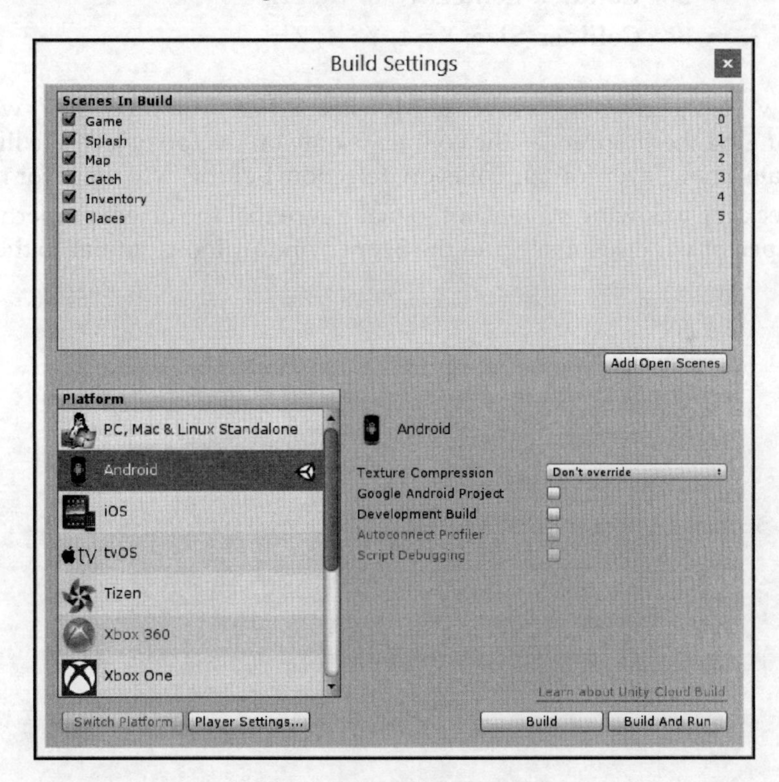

Build Settings with Places scene added

If you run the game in the editor, unfortunately, you will still be unable to click on a place to open the `Places` scene. That is because we need to allow our `PlaceMarker` to act as a collider, something we omitted in the last chapter. Perform the following instructions to add the collider and other settings to the `PlaceMarker`:

1. Drag the `PlaceMarker` prefab we constructed in the previous chapter from the `Assets/FoodyGo/Prefabs` folder into the **Hierarchy** window. Note that we are altering the `PlaceMarker` and not the `PlacesMarker` prefab we modified in this chapter.
2. Select the `PlaceMarker` in the **Hierarchy** window, and then from the menu, select **Component** | **Physics** | **Box Collider**.
3. Double-click on the `PlaceMarker` to focus the object in the **Scene** window. You will see a green box at the base of the object; this is the collider.
4. In the **Inspector** window, change the **Box Collider** – **Center** and **Size** to the values shown in the following list:
    - **Box Collider**: **Center**: X = `0`, Y = `2`, Z = `0`
    - **Box Collider**: **Size**: X = `1`, Y = `1`, Z = `.2`
5. Now, change the objects layer to **Monster**. When prompted, don't worry about changing the children, as the collider is only on the top object. We likely should create a new layer for our collision detection, but this will work for now.
6. Check the following screenshot to make sure that the object properties in the **Inspector** window match and the **Scene** window looks similar to the following:

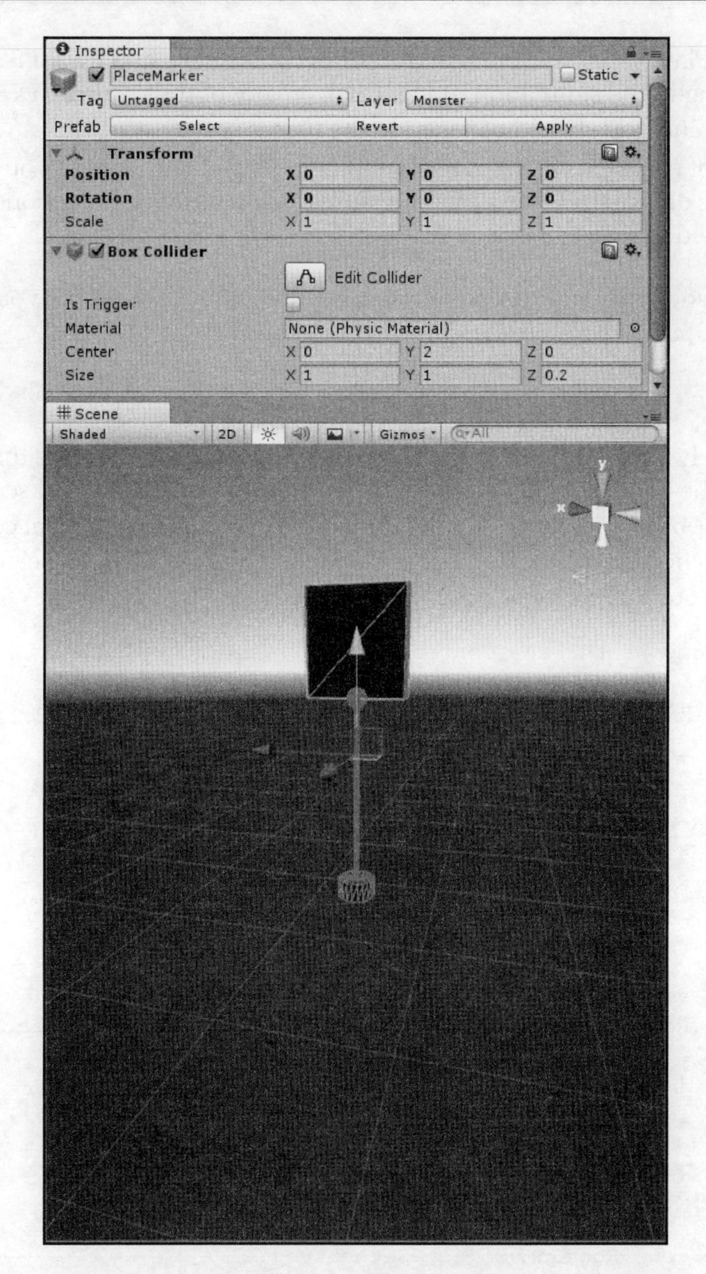

Box Collider settings, as shown in Scene window

7. Click on the **Apply** button beside the **Prefab** options in the **Inspector** window to save the changes. Then, delete the `PlaceMarker` from the **Hierarchy** window.

8. Press Play to run the game in the editor. Make sure that the GPS service is running in the simulation mode. Click on any of the `PlaceMarker` signs in the scene, and you should be taken to the Places scene.
9. Also, of course, as always, ensure that you build and deploy your game to your mobile device for testing, making sure that everything transitions and looks as expected.

If you encounter problems running the game, ensure that you consult `Chapter 10`, *Troubleshooting*, for help.

That was relatively simple to set up, so let's take a quick look at the code that wires up the collision. Open the `GameManager` script in the editor of your choice, and scroll down to the `HandleHitGameObject` method. At the top of the method is our original code that handles when the player clicks on a monster and is then taken to the `Catch` scene. Just below that, we handle when our `PlaceMarker` gets hit (selected):

```
if (go.GetComponent<PlacesController>() != null)
{
        print("Places hit, need to open places scene ");
        //check if the scene has already been run
        if (PlacesScene == null)
        {
             SceneManager.LoadSceneAsync(PlacesSceneName,
LoadSceneMode.Additive);
        }
        else
        {
             //the scene has run before, reactivate it
             var psc =
PlacesScene.RootGameObject.GetComponent<PlacesSceneController>();
             if (psc != null)
             {
                 var pc = go.GetComponent<PlacesController>();
                 psc.ResetScene(pc.placeId, pc.location);
             }
             PlacesScene.RootGameObject.SetActive(true);
         }
         MapScene.RootGameObject.SetActive(false);
  }
```

This block of code is very similar to what we used when the monster object (`MonsterController`) was hit or selected. If you follow the comments, things are quite self-explanatory. As you can see, following this same pattern we could easily add other objects you may want to interact with quickly.

You may also notice that it doesn't matter how far the places are away from the player, they can still interact with them. This is intentional at this point, but it could be something very easy to fix by just setting the length of the ray when testing for collisions. If you scroll up in the code to the `RegisterHitGameObject` method, you will see the ray is set to infinity:

```
public bool RegisterHitGameObject(PointerEventData data)
{
    int mask = BuildLayerMask();
    Ray ray = Camera.main.ScreenPointToRay(data.position);
    RaycastHit hitInfo;
    if (Physics.Raycast(ray, out hitInfo, Mathf.Infinity, mask))
      {
        print("Object hit " + hitInfo.collider.gameObject.name);
        var go = hitInfo.collider.gameObject;
        HandleHitGameObject(go);

        return true;
      }
      return false;
}
```

You could change this value to a hardcoded value, such as 100 or perhaps revisit our discussion on GPS accuracy and set this value based on the current accuracy of the device. In either case, this is another game mechanic that can be easily set, but can have wide variations to how your game will be played.

Summary

This has been an intense chapter dedicated to building the Places scene as a venue for the player to interact with the game world mixed with some of the virtual world. We started by building the foundation of the scene. Then, we moved on to incorporating the Google Street View images as our scene backdrop in order to provide a deeper sense of real-world interaction with the game. From there, we enhanced the marker sign with photo slideshow provided by Google Places Photos and other elements, such as the name, rating, and price. Now that we gathered detailed information about the place, we added the buttons for interaction and dialogs. This gave us some extensive experience working with the Unity uGUI elements, including using layouts to stack elements. After getting the interactive elements in place, we moved on to understanding how the game mechanics of selling worked. That led us to upgrade the inventory database with new tables and CRUD methods for inventory items and the player. We then connected all the pieces together and added the Places scene to the game. Then, to get everything working together, we added a box collider to the marker so that it could be interacted with by the player in the **Map** scene.

We are getting close to finishing up the game, and the next chapter will take a diversion into adding multiplayer networking support for the game. Also, we will spend some additional time enhancing the game with particle and visual effects.

9
Finishing the Game

At the end of the previous chapter, we wrapped up most of the major features we planned to include in the game. As much as we could continue the final chapters of this book working on every little detail of the game, that would become quite repetitive and we likely would lose several readers. The game is, after all, only meant to be an amusing demonstration of the concepts of building a location-based AR game. Hopefully, you pondered your own unique game design and how that could be developed as you worked your way through this book. So, instead of building on the Foody GO game, this chapter will go over the following things:

- Outstanding development tasks
- Missing development skills
- Cleaning up assets
- Releasing the game
- Problems with building a location-based game
- Location-based multiplayer games
- Firebase as a multiplayer platform
- Other location-based game ideas
- The future of the genre

Outstanding development tasks

Having completed development on the Foody Go demo, it will be helpful to review the outstanding tasks a developer may undertake to complete the game for commercial release. This exercise will certainly be beneficial for any new developers planning to release a full commercial game in the future.

You may have heard of the *80/20* rule, which states that it will take *20%* effort to complete a task to *80%*, and another *80%* effort to complete the last *20%*. This rule applies well to most tasks, but especially well to software and game development. If you consider that the demo game is *80%* complete, it will therefore take another four times more effort (6 chapters of development * 4 = 24 chapters) to finish the entire game. This may seem like a huge amount of work considering everything that has been completed, but let's summarize by scene what items are outstanding:

- **Map** scene:
  - Sound effects or music
  - Skybox with clouds and sun or night, depending on the time of day
  - Improved monster spawning, central server
  - Visual shader effects
  - Map styles
- **Catch** scene:
  - Sound effects and music
  - Visual shader effects
  - Gyro camera
  - Option to enable/disable AR (background camera)
  - Switching freeze balls from inventory
  - Flee the scene
  - Take a picture with the device camera
- **Inventory** (**Home**) scene:
  - Item inventory
  - Monster details
  - Character details, stats, and level
  - Monster index, if more monsters or other creatures are developed
- **Places** scene:
  - Sound effects and music
  - Visual shader effects
  - Gyro camera
  - Option to enable/disable AR or the Street View backdrop
  - Place tracks the chef(s) at a location
  - Monster animations
  - Improved UI
- **Splash** scene:
  - Sound or music

        - Loading effects
        - Images
- **Game** scene:
    - Option to choose/customize character (male/female)
    - Varied monsters, character development
    - Sound service or manager
    - **Massively Multiplayer Online** (**MMO**) optional
    - Asset cleanup
    - Bug fixes (all scenes)

Aside from the multiplayer item, which we will talk about later, the list includes numerous items that are not overly complex, but still require several additional hours of effort to complete. Visual effects (shaders), styles, and bug fixes are especially problematic and can literally be a black hole to development time. Developers have been known to spend several months or more working on specific shaders just to get a specific look right for their game. Hopefully, you can now better appreciate how the *80/20* rule certainly will apply to our demo game.

Of course, feel free to take the demo game we have here and complete it as your own, with your own features and game mechanics, as a location-based AR game. However, it is important to be mindful of the effort you are spending on developing or completing the game, especially if it is your first game. Be sure to set yourself a release date and then do your best to publish the game as an alpha/beta/release on that date. This will make sure that your game is reaching out to players in a timely fashion and you will get immediate feedback of some form. Following this strategy not only makes you more aware of deadlines, but also reinforces your estimation of effort as it relates to reward (feedback).

As an example of how easy it is to spend time working with or exploring visual effects, follow this exercise:

1. Open up any scene in Unity, either from the demo game we just built or another scene of your choice.
2. From the menu, select **Assets** | **Import Package** | **Effects**. This will import the `Unity Standard Assets Effects` package. Continue with the import as you did earlier.
3. Locate the `Main Camera` in your scene. If you are unsure where it is, use the search feature at the top of the **Hierarchy** window. Just start typing `main camera` or `camera` in the search area and your scene camera will be shown in the hierarchy.

It is easy to get carried away with using filters and other visuals. Just be aware that these filters may have a direct and significant impact on performance, memory, and game play. Use these in your finished game with caution.

4. Select the camera and then, from the menu, select **Component** | **Bloom & Glow** | **Bloom**. This will add a `Bloom` filter to your camera and the **Game** window will take on the appearance of a brighter hue.
5. Feel free to continue adding various effects (filters) to your camera, every so often running the scene to see what effect the filter has. It is likely that, after a quick couple of hours pass by, you will realize that there are a lot of choices and options here. The following is an example from the **Map** scene in the Foody Go game:

Before and after screenshots showing various image effects used on camera (note clouds skybox was added earlier)

6. Try not only adding various effects, but also try moving the order of the effect components on the camera itself. The following is an image showing the camera image effects that were used in the example scene:

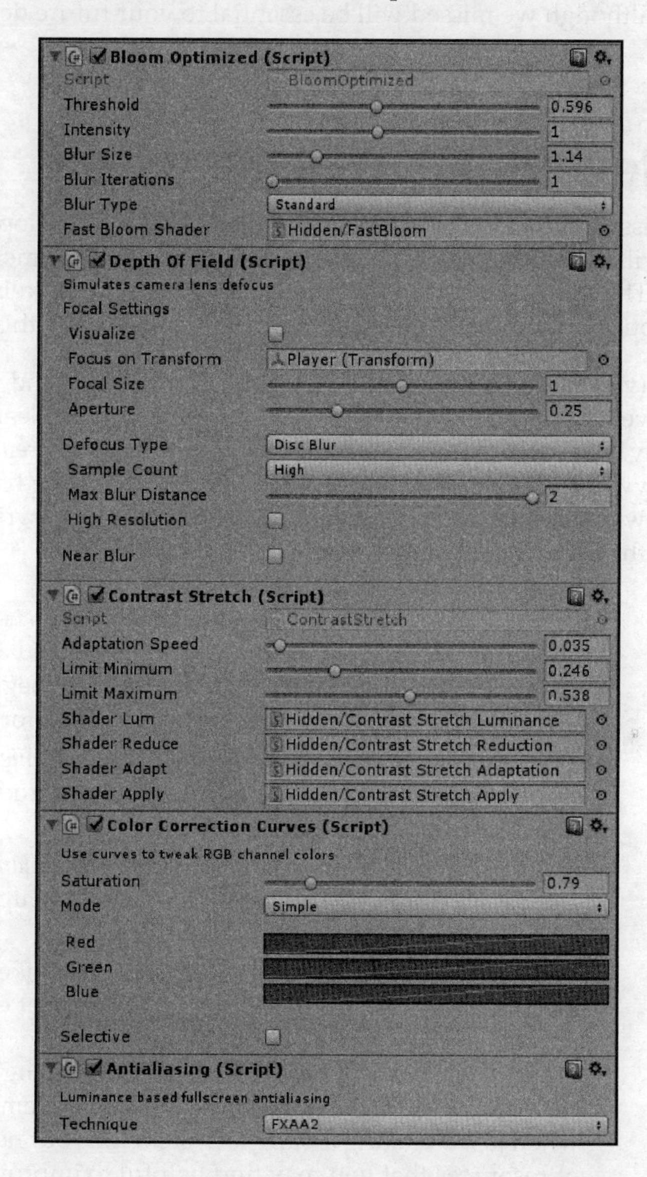

Camera image effects used in previous scene

The outstanding task lists identified some skillsets that we barely covered or didn't cover at all in the book. While it was intentional since those skills were not critical to the games' functional elements we were building, they are still critical to any game. We will review those skills, which although we missed will be essential to your future development efforts, in the next section.

Missing development skills

Even though we missed or barely touched some of these areas of development for our demo game, they will indeed be critical to completing a stunning commercial or just cool game in the genre. The following is a prioritized list of development skills, with a sublist of resources, which you should direct your attention to after completing this book:

- **Shaders** (visual effects and lighting): This is a broad topic and you would likely never have to write a shader, but you will certainly have to learn how to use them effectively. Shaders are to game development like air is to breathing; essential and everywhere. Learning to develop shaders is an advanced skill, but even a little knowledge can go a long way. Even if you plan not to write your own shaders, the following resources are worth taking a look at:
    - *WikiBooks*: (`https://en.wikibooks.org/wiki/Cg_Programming/Unity`), has a good introduction to shader programming and a good set of resources. The site may be overwhelming to beginners or those that like to follow a learning path, in a good book for example.
    - *Mastering Unity Shader and Effects* (2016) by *Jamie Dean, Packt Publishing* is an excellent book and a good introduction for those new to shader development.
    - *Unity 5.x Shaders and Effects Cookbook* (2016), *Alan Zucconi, Kenny Lammers, Packt Publishing* is an excellent, although more advanced book. Check this one out last.
    - **Particle effects**: Again, this is a skill you can likely get by without, but there will come a time when you want that extra special-particle effect, which will require knowledge of building or modifying a custom-particle effect. Even having knowledge of how the various settings on the Unity particle system may be helpful when just tweaking a particle system for your needs. Here is a list of resources that you may find helpful to improve your knowledge of the Unity particle system:

    - *Unity Learning* (`https://unity3d.com/learn/tutorials/topics/graphics/particle-system`), is a great introduction to use the particle effects system
  - *Mastering Unity Shader and Effects* (2016) by *Jamie Dean, Packt Publishing,* has a good section on developing shaders for particle effects

- **Multiplayer** (networking): Of course, if you want to convert the demo game to an **Massively Multiplayer Online** (**MMO**) game, then this skill will be essential. Unfortunately though, the required specific skills and knowledge in this area will often depend on the networking infrastructure your game depends on. We will get into why this is so and talk about various networking solutions and resources later in this chapter.
- **Animation** (Mecanim): Many developers will automatically assume animation and Mecanim is just for animating characters or humanoids. While that certainly is true, you can accomplish several things with Mecanim animations, including of course character animation. Here are some great resources on animation in Unity:
  - *Unity Learning,* (`https://unity3d.com/learn/tutorials/topics/animation/animate-anything-mecanim`) is a good quick tutorial on using Mecanim.
  - *Unity Animation Essentials* (2015) by *Alan Thorn, Packt Publishing,* is a good short introduction to animation.
  - *Unity Character Animation with Mecanim* (2015) by *Jamie Dean, Packt Publishing,* is another excellent book by Jamie. There is a lot of good material in this book.
- **Audio**: This is one of those often overlooked skills that are so essential to good games. Of course, the other issue is that there is very little information out there about the best practices of adding sound and music to a Unity game. Sounds like this would be another good topic for a book, but until then, here are a couple of resources:
  - *Unity Learning* (`https://unity3d.com/learn/tutorials/topics/audio/audio-listeners-sources?playlist=17096`) is a good place to start
  - *Unity Learning* (`https://unity3d.com/learn/tutorials/topics/audio/sound-effects-unity-5?playlist=17096`) is a good discussion on sound effects

- **Texturing**: This is the art of overlaying 2D images onto a 3D model. It is a skill that falls outside development, but if you are a small indie developer working on your own game, some knowledge in this area will be beneficial. This skill aligns itself well with or as a part of 3D modeling. It is also an extension of shaders and shader programming. Here is a good starting point on physically-based substances (textures/materials) with Unity:
    - *Unity Learning* (`https://unity3d.com/learn/tutorials/modules/intermediate/graphics/substance/introduction`) is a rather advanced tutorial series using Substance designer from Allegorithmic, but it is certainly worth taking a look at by anyone interested in getting insights into the next level of game development
- **3D modeling** (character development): This is another skill game developers could do without and still do well. There is certainly enough 3D content out there. However, learning just the workflow of modeling a 3D object or character can be another great addition to your skill set. Not to mention, if you are indie developer, at some point you will be opening a 3D modeling program to build a custom model. Of course, some of the skills you learn in this area will be specific to the software you are using. Here are a couple of software-specific packages with tutorials that will help you get started:
    - **iClone Character Creator** (Reallusion), `https://www.reallusion.com/iclone/game/` , is an excellent software package with a good selection of tutorial videos that will allow you to easily create custom characters. If you forgot, the characters used in this game are iClone characters provided free from Reallusion.
    - **Blender**, `https://www.blender.org/`,is the standard free tool for modeling 3D content. It does require an investment in time to get familiar with this tool.

Of course, if you don't have time to spend improving skills, you can always look for an asset in the Unity Asset Store for complete functionality. We will discuss several guidelines on selecting and using assets in the next section.

Cleaning up assets

When you first start developing with Unity, the Asset Store can be both a blessing and a curse. There are several excellent assets available in the store that can give you great strides in developing a game. However, there is often a price that comes with using too many assets, and assets that come with additional baggage like demo scenes or other packages. We discussed asset packages earlier, but here are some guidelines that will help you decide how and when to buy and/or download assets:

- Ask yourself why you need the asset:
    - It would take too long to write your own: If you consider that most assets are priced at less than $100, it can be a compelling reason to buy, especially if the asset does a lot of things you need and you value your time. Just be sure that you need the asset, even if it's free, by following the other guidelines in this list.
    - It provides content outside your expertise: Probably, the best reason to pick up an asset is if you are new to Unity and/or game development. Ensure that you follow the other guidelines in this list, just to confirm that it makes sense to grab it.
    - It is on sale: Avoid this trap; this should be a secondary reason that perhaps makes the decision for you.
    - It has great reviews: Reviews are great, but understand the core reason why you need the asset. Even if the asset is an editor tool be mindful that it still takes up space in your project.
    - Other cool games use it: This is another trap; unless you are attempting to exactly match functionality and infrastructure, ensure that other reasons make sense.
- Determine what support exists for the asset:
    - The asset version is compatible with your Unity version. This is a must with newcomers, as dealing with version conflicts can be a nightmare. For more experienced developers, this may not be an issue.
    - The developer supports updates to the asset. Check the asset page and make sure that the developer supports the asset. In some cases, this may be less of an issue, for artistic assets for instance. For assets with source code, this becomes less of an issue. However, high-performance or tightly integrated assets like water, terrain, and animation are something you want a developer to provide frequent updates for.

    - The reviews are favorable and current. Be sure that the asset has reviews and they are current and, of course, positive. Again, this is less of an issue for free assets because developers won't typically comment on free assets. For other assets, this can be critical. If there is any hint of issues with an asset, it is often better to avoid it.
    - The asset has a Unity forum thread. This is often a great indication that the developer is responsive to comments and bug fixes. Take a read through the posts and check the dates and type of questions. This can often give you some great insight into how the asset works and if it really will work for your project.
    - The asset provides free documentation. This is often a given now that most assets will provide a direct link to their documentation, whether you buy or not. Do a quick read through and determine if you can follow documentation and if it is something that will fit your project.
- Review the asset package contents:
    - The asset contains plugin folders. This typically means that the asset will contain compiled libraries that may require special configuration when deploying the game. Of course, this could be an issue and something to be very aware of. Some developers may preview only the assets where they have access to all the source code. Yet this can be a trade-off sometimes, because the asset may perform dramatically better as a compiled plugin.
    - The asset contains extra content. An asset that contains additional demo files can be a great learning resource. Some assets may add the Unity Standard Asset packages or other content that could conflict with your existing assets. This may not be a reason to avoid using an asset, but it is helpful to be aware before you pull an asset into your project. It is also often more beneficial to bring an asset with a lot of demos into a test project, where you can strip down the asset to only the bare essentials and then export it for use in your project.
    - The asset content structure is well organized. Try to avoid assets that load content into multiple root folders. Not only can this be a hindrance to asset management, but it is often a sign that things that don't need to be loaded might be getting loaded'.
    - The asset content is designed for your deployment platform. This is a big one; don't expect to use content designed for a desktop on a mobile device. If the asset does not have content designed for your platform, then keep looking.

  - Asset scripts are in your preferred language. This isn't always a given and you don't want it to be a surprise. If you prefer C#, for instance, then just ensure that all the content script files end with `.cs`.
- Compare the alternatives:
  - The asset has a lot of competition. This is generally a good thing for the developer, since it often indicates that the asset will be priced well for the function it delivers. It, also, most likely means that building the asset by yourself would not be worth the time. However, a competitive market can make it overwhelming to determine which asset is right for your project as there may be overlapping features and functionalities.
  - The asset has no competition. There are a few assets on the store that are so well done; they require such advanced knowledge of Unity and game development, nobody bothers to compete with them. Alternatively, the asset may be a new development concept and they are the first to release. As long as you follow the other guidelines, you can determine if the asset is a good pick.
- Other considerations:
  - The asset is compatible with your target platforms. Make absolutely sure that the asset will work with your deployment platforms. In many cases, this can be a non-starter.
  - The asset has a free version. Anytime you see a free version of an asset, it certainly helps ease your decision. Of course, you do still need to download the asset and do all the configuration and setup for your project. Then, if you find that the asset won't work, go through the problem of removing the contents later. Be wary of free versions, in the end they may just be a waste of time.
  - The asset is a game starter pack or large `framework.Starter packs` and frameworks can be great, but you need to be especially careful when going down this road. Of course, this book itself develops a starter location-based AR game. Yet, in reading this book, you are quite familiar with all the nuances of the project. This may not be the case if you downloaded a completely foreign starter kit. If you plan to build your own game off a starter, do your homework and ensure that you understand the nuances of the kit.

Now that we reviewed an extensive set of guidelines gather assets for your project, we also want to cover how to clean up and remove the additional asset content that your project doesn't need. The following is a list of options to manage assets followed by a practical tip on how to easily do a basic project clean up:

- Block unneeded content on import. If you did your homework well enough, you should have a good idea of all the items the asset is going to want to import. It is also a good practice to always import the project into a blank or test project. This should give you an idea of what content you need and what may just be used for demos. Then, when you do the actual import into your working project, you can uncheck the content you don't need.
- A+ Assets Explorer, the paid version of this tool is reasonable and provides a good set of tools for projects with a large amount of assets, which could include most projects. This tool won't clean up assets, it will just tell you where your problems are.
- Don't relocate imported asset content. Never relocate asset content that could be updated by the asset developer later. Moving content around will only cause problems later; when you upgrade, files may get duplicated or dropped in the wrong folders. On the other hand, if you know the content won't be upgraded, then moving it around would not be an issue. Of course, this rule also does not apply to any content that you may import manually.
- Do a full project asset export. When your project has reached a stage where you feel it is time for a clean or you just want to remove unnecessary assets, then do a full asset export cleanup, as follows:
    1. Open your game in the Unity editor and ensure that the current scene and project are saved.
    2. Ensure that all the custom game content (scripts, prefabs, scenes, materials, and images) is in a root project folder.

If we use the **FoodyGo** project as an example, you would create a new `Scenes` folder under the `FoodyGo` folder and drag all the scenes from the `Asset` folder to the new `Assets/FoodyGo/Scenes` folder. Then, all the custom content would be under the `Assets/FoodyGo` folder.

3. Select the top level custom content folder and select **Assets** | **Export Package...** from the menu, which will open the **Exporting package** dialog. Ensure that the **Include dependencies** checkbox is checked at the bottom of the form and as shown in the sample dialog, as follows:

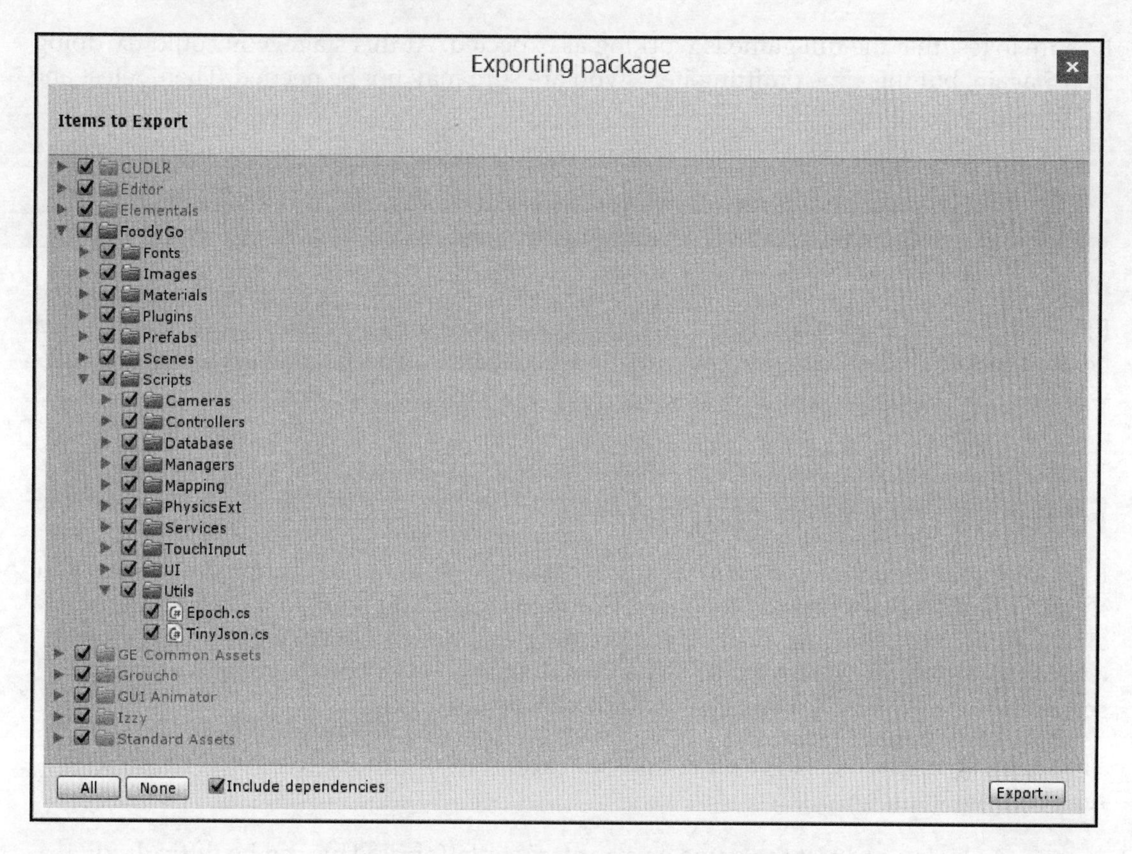

Exporting custom and dependent content

4. Click on the **Export...** button of the dialog to open the **Export package...** dialog. Choose a good location to save your package, name it something memorable, and click on **Save**.
5. Open a new instance of Unity with a clean empty project. Now, from the menu, select **Assets** | **Import Package** | **Custom Package...** and use the open dialog to select the file you just saved. Then, click on **Open** to start import of the package.

6. After the **Import package** dialog opens, confirm that all the custom and some of the dependent content is there. If you didn't clean your project earlier, then the content will be measurably reduced. Click on the Import button to import the package.

Be sure to test that the full game is working as expected. At this stage, you could try doing step 5 again, but this time omitting items you are sure may not be needed. Then, retest and try from step 5 again.

There are, of course, other tools and methods to reduce asset bloat. This method, however, allows you to test dependencies in isolation and at the same time, creates a backup of your game.

Effectively managing the assets in your project is a must for any game that has progressed beyond concept. Cleaning and removing unused assets is especially an exercise you will want to cover before releasing your game, and is discussed in the next section.

Releasing the game

Even if you are developing a game as a hobby, for recreation or educational purposes you should release it at some point to others. Getting feedback on your game can be an exhilarating, humbling and even a disappointing experience. Don't be discouraged by criticism and pay attention to your tester's comments good or bad. Sending someone a screenshot or movie clip of your game for review is not enough.

The following is another set of guidelines that will improve the success of your game release, even if it is only to friends:

- Deploy and test the game on your target platforms. This can be difficult for a platform like Android, where there are so many variations of device. However, just testing on a single Android device could reveal multiple issues your players will encounter.
- Test with a small target audience, it is important that your testers match the audience you are releasing your game to. If you don't have any testers like that, then you need to go out and find some. Fortunately, there are numerous game development sites and forums where you can find very honest test candidates.

- Fix the bugs, this is typically the point where that initial tester feedback excites you to further complete your game or forget the project entirely. Just remember that criticism is good and don't be afraid to consider suggestions. Assess the project honestly and consider if you need to go back to development to fix any serious or critical bugs.

Critical bugs are generally defined as unexpected game crashing or data loss. Serious bugs are something that interferes with game play or hinders the player in a bad way.

- Release to a larger target audience. If possible, do another release to a larger target audience. This may not always be practical and in some cases, it may just be better to do an actual release to a store.
- Release to an App Store. Depending on your target platform, this may or may not be a huge hurdle. Even if you release the game at an entry level price or free, this will be a great experience.
- Automate the process. If you are happy with your experience of deploying to the app store, then at some point you will want to update the game with new features and bug fixes. In order to minimize errors and delays in putting out updates, you will want to automate your release process. Automating the release will not only make you more responsive to change, but also free up your time for bug fixes and new features.

Now that we have discussed the skills, assets, and release guidelines, it is time to get into the dirty secrets of location-based games. In the next section, we will talk about the difficulties of building a location-based game.

Problems with location-based games

Location-based games are a relatively new genre in games brought on by the prolific use of GPS in mobile devices. In many ways, this genre is still cutting edge and developers are still finding their way with what works with players. Let's take a look at the main problems with location-based games and what options you may or may not have getting around them:

- **GIS mapping services**: Likely at the core of your game will be the requirement to display location on a map. The following is a list of options for how you could perform that:

- **Use your own GIS servers**: While this is an option and there are a few good open source GIS platforms out there, be aware that you will still need to support the servers and GIS platforms are notoriously CPU hungry.
- **Use Google Static Maps**: This is the option we used in this book with good success. However, be aware that using the Google Maps Static API is subject to limitations. The limitation is 2,500 requests per day per IP address.
- **Use Google Maps**: Google Maps is currently free on Android or iOS platforms, provided you use the provided SDK. Perhaps in the future, this may also be an option for Unity developers out of the box.
- **Use a different service**: There are a number of other free GIS services out there. This may be a valid option if you are okay with the map styles. Alternatively, a shader could be used to alter the visual style in some manner that is more aesthetic to your game.
- **Location data**: Location-based games are, by their nature, tied to location data, be it real world or virtual. The following list reviews a few options for accessing location data:
    - **Use you own GIS servers**: Again, this is an expensive option but it certainly has its advantages. You could, for instance, spawn monsters or creatures according to advanced GIS rules. Taking this step will require some advanced knowledge of GIS that is far outside the scope of this book.
    - **Use Google Places**: The option we used in this book worked well for our small example but, unfortunately, may not scale well. Remember, those usage limits were quite restrictive. However, you can also procure licensing from Google to increase limits at a price. So, it may be a viable option if your game is earning revenue.
    - **Use another service**: There are other location-based data services out there (Foursquare, for example) that provide similar or, in some cases, better data depending on your needs and region. This could be a viable option and you could use some of the skills you picked up in this book to connect to those other services.
- **Multiplayer support**: Unlike other genre's, the location-based game cannot just expect to use off-the-shelf multiplayer services. Location-based games are continuous and players often play over extended periods of time. As well, players should only interact with each other at a certain physical distance. We will get into more details about multiplayer networking in the next section, but here is a list of multiplayer options:

- **Photon PUN**: Photon is a great multiplayer service and is easy to set up and get running quickly. However, like other multiplayer networking services, there is only limited support for extended state transition. This means that players who connect after being away may be overwhelmed with update messages.
- **Unity UNET**. The Unity multiplayer networking system, UNET, is a robust and useful framework for peer to peer games. For games that require extended state and regional filters, UNET is certainly not it.
- **Other multiplayer platforms**: There are plenty of other options around that may work. The key thing to keep in mind is state management across sessions and limiting player interaction to geographic regions. The best options here are platforms that provide you with a server that you can customize according to your needs.
- **Develop your own server**: If you have already taken the serious jump of providing your own GIS data services, this is certainly an option. We will explore this concept a little further in the next section.
- **Use an online real-time cloud database**: This may sound like an out-of-the-box and whacky solution, yet this is a viable option and one that we will seriously explore in the next section on multiplayer networking.

Hopefully, this honest discussion of the problems with developing a location-based game hasn't scared you away from continuing. As we discussed at the beginning of the book though, location-based games come with a number of unique and difficult problems, with supporting multiplayer being at the top of the list, but in the next section, we will discuss some strategies to develop a location-based multiplayer game.

Location-based multiplayer game

Adding multiplayer support to a location-based game, as discussed in the preceding section, will not be easy. In fact, we avoided multiplayer support for our demo game because of the added complexity. Also, it was important to demonstrate that you could still build a functional location-based game without backend servers and multiplayer. In the end though, you are now here wondering how to add multiplayer support to your game.

In the preceding section, we discussed a couple of viable options to add multiplayer support: developing your own server, extending an existing platform and using an online real-time cloud database. These options may be attractive, but before we get into the specifics, let's review the fundamental problems of location-based multiplayer games, as listed:

- **Game is continuous**: Our game needs to continually save state globally for all players. If you don't appreciate the magnitude of that last sentence, take a second and think about it. When players reconnect, the whole world around them needs to be updated almost instantly, regardless of their last location. Depending on the amount of state you are saving, this could be a difficult problem. This is why we depended on those backend Google services for our game locations.
- **Players only need to interact locally, if at all**: Location-based games have received a lot of criticism and concern for allowing players to interact with each other, especially for games that target younger audiences. Therefore, you will generally only want to allow your players to interact through places, stores, or other virtual constructs regionally. An example of a regional interaction may be setting a lure on a location for instance. All players would be able to benefit from the lure, but not directly interact with each other.
- **Game state needs regional filters**: Players should only be able to see or interact with the world within their map area; even our ability to search the Google Places service from a location and radius needed to be further filtered. Ideally then, the game state a player needs to pull should support some form of regional searches. This can be tricky, especially if your experience with GIS is limited. Fortunately, we will identify a solution for this shortly.

Now that we understand the difficulties each of the solutions face, let's review the list of features each solution will need to address in the following table:

Features/Requirements	**Develop your own server**	**Extend the existing multiplayer server**	**Real-time cloud based database**
Security (Access)	Need to develop your own access and user database.	Use or customize the existing access	Provides a robust secure platform with multiple options for user access

Security (Data)	Need to develop your own security mechanism	Likely supports data protection and player cheating	May need to customize database structure to support data security
Game State	Require a backend database to persist state	Likely already implements a database	Is a database
Continuous game state	Full control over database	Want a server that supports MMO or MMORPG games	Full control over database
Isolating player interaction by region	Use `geohash*` to isolate players by region	Would likely need to customize the world to use regions and/or `geohash*`	Use `geohash*` to isolate players by region
Limit game state updates by region	Managed by `geohash*`	Managed by custom `geohash*`	Managed by `geohash*`
Scalability: Adding players	Need to manage additional servers and infrastructure	Need to manage additional servers and infrastructure, possibly extra license fees	Cloud solution made to be extensible
Extensibility: Adding game features	Development, then rolling out updates to servers and clients	Development and then rolling out features to server and clients	Can easily version database and then update clients as required
Data backups	Need to support this on your own	Likely need to support this on your own	Likely supports database backups, but only at a certain level
Infrastructure (servers)	Need at least one server running all the time	Need at least one server running all the time	Cloud supported

Price	Inexpensive or free, depending on the available infrastructure (servers)	Price of multiplayer software plus infrastructure costs (server)	Likely free to start with

* `geohash` will be explained further on. **Geohash** is a method of hierarchically representing spatial data on a grid using a unique ID that is a sequence of characters. This is likely difficult to comprehend without an example, so the following diagram may help to explain this visually:

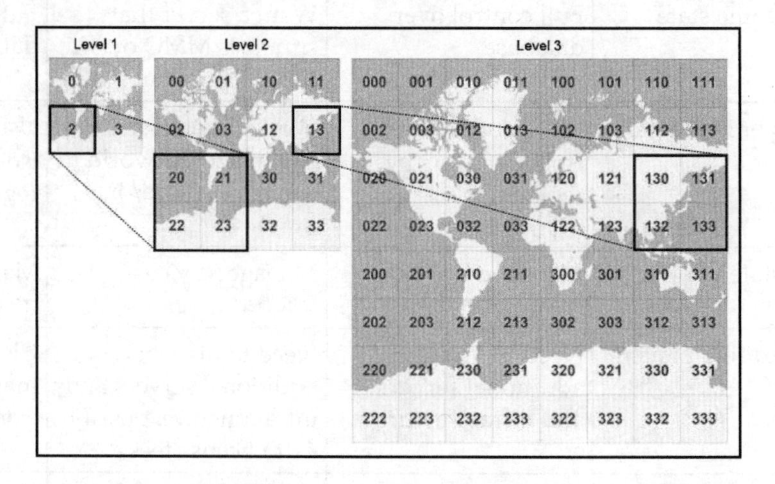

Diagram showing how geohash subdivides the world at various levels

The grid shown in the preceding diagram is more representative of a tile geohash, whereas we would likely use the more standard Geohash-36, which just means each level is subdivided into 36 sections. Check out `http://geohash.gofreerange.com/` if you want to see what the geohash for your address or city is.

Once we can isolate a location or region to just its geohash representation, this allows us to do very fast and efficient ordering of messages by region. For instance, if we know a player is at geohash *9q8y* (San Francisco area, CA), then we could isolate all messages that started with that geohash. Therefore, if another player at Alcatraz Island has a geohash of *9q8znn* and because the first four characters are different, they would not view activity from the other.

The following is a screenshot from `http://geohash.gofreerange.com/` that shows this visually:

Showing geohash levels of the San Francisco area

If you look at the map diagram closely, you can see how the geohash grid is further subdivided over Alcatraz. You can actually get down to almost a location point (geohash level 12), but you will likely never need to use that level. In the example, we are using geohash level 4, but that doesn't mean you could use a higher or lower geohash level to control your regional messaging.

A `geohash-36` library has been included with the FoodyGo final source code in the `Chapter 9 Assets` folder included in the downloaded source code.

As you can see by the preceding table, things generally point to using a real-time cloud database. Of course, you may still have your own bias towards a particular solution but hopefully, you better understand some of the pitfalls of each. That table does try to address the high level requirements but software development is always about the details.

There are a number of real-time cloud-based databases in the market, but only a couple that support direct integration with Unity. First, there is **Firebase** from Google, which is an app platform that supports real-time database and many other features like analytics, ads, crash reporting, hosting, storage and more.

Second, at the time of this writing, Unity is developing a real-time database that will likely support all the same features as Firebase. Unfortunately, Unity seems to always be one step behind its competitors and that option is currently in closed alpha, which means we will cover how you could use Firebase as a multiplayer platform in the next section.

Firebase as a multiplayer platform

For many developers, using a cloud-based real-time database may be way out there and not even worth consideration. Of course, it also very much depends on the type of game you are developing. You likely wouldn't want to use a real-time database for a FPS, but fortunately, Unity has excellent options for that. For a location-based game, though, a real-time database makes ideal sense for all the reasons we mentioned earlier.

Even if you are not making a location-based game and one that doesn't require ultra-high performance updates, you should consider Firebase. The free version of Firebase real-time allows 100 concurrent users and the first level subscription provides for unlimited concurrent users. Other multiplayer platforms generally start charging after 20 concurrent users.

We won't add Firebase as a multiplayer platform to the sample game we developed over the course of the book. That would likely need several chapters or even a book on its own to cover the numerous details. Instead we will look at setting up and using one of the Firebase samples. First, follow the directions given to download the sample Firebase SDK database project:

1. Open a web browser and navigate to the following URL: `https://github.com/firebase/quickstart-unity`.
2. Follow the directions on the GitHub page to either download the SDK samples as a ZIP file or using Git to clone the repository to your computer. If you downloaded the samples as a ZIP file, unzip them into a folder you can locate later.
3. Start a new instance of Unity and click on the **Open** button at the top of the project selection window. Navigate to the folder you unzipped or cloned the SDK samples to. Then, open the `database` folder and select the `testapp` folder as the Unity project folder. Click on the **Open** button to load the project.
4. After the project is open, you will see some compile errors. This is normal and nothing to worry about. We will resolve these shortly.
5. Open **MainScene** by double-clicking on the **MainScene** scene in the `Assets/TestApp` folder in the **Project** window. The scene will load blank.

If Unity is still set to use the mobile layout, you may want to change the back to **Default** layout for the rest of this exercise. From the menu, select **Window | Layouts | Default**.

With the database sample project set up and ready to go, now we need to create a Firebase account and perform some configuration:

1. Use your web browser to go to the Firebase site at `https://firebase.google.com/`.
2. Near the top of this page will be a large button with the text, **Get Started for Free**. Click on this button and, when prompted, log in with your Google account, preferably the same one you used to create the Google Places API key earlier.
3. After you log in, you will be taken to the **Console** and prompted to import or create a new project. For now, we will just create a new project.
4. You will be prompted to name you project and select a region. Name your project `TestApp` and then, choose a region from the list that matches your location, as shown:

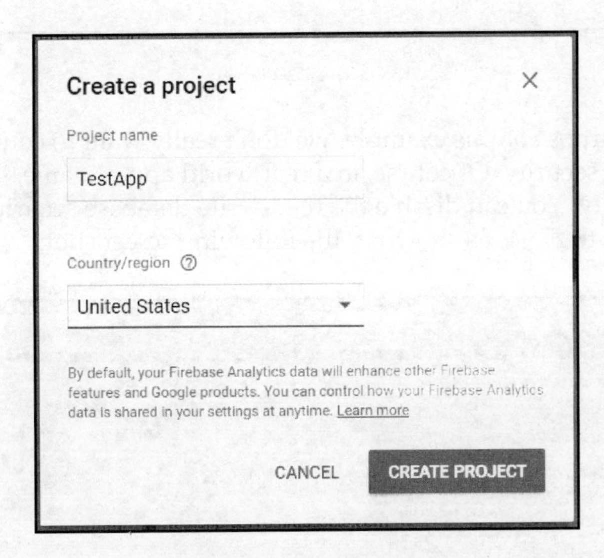

The Firebase create project window

5. Then, click on the **Create Project** button to create the new project. After the project is created, you will be taken to the Firebase Console for the project.

6. On the left-hand side panel, select the **Database** item to open the **Real-time Database** panel. Then, in the panel, select the **Rules** tab at the top. The following screenshot is shown for reference:

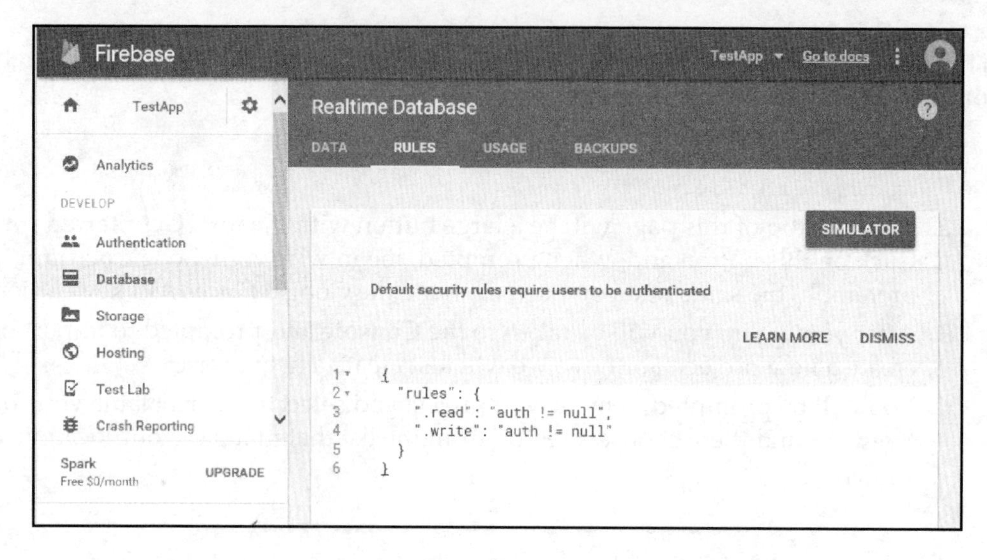

Database read/write rules

7. For the purpose of this example, we don't really want to concern ourselves with setting up security. Of course, in a real-world app or game, security should be the first priority. You can disable the read/write database security by setting each of the values to `true`, as shown in the following screenshot:

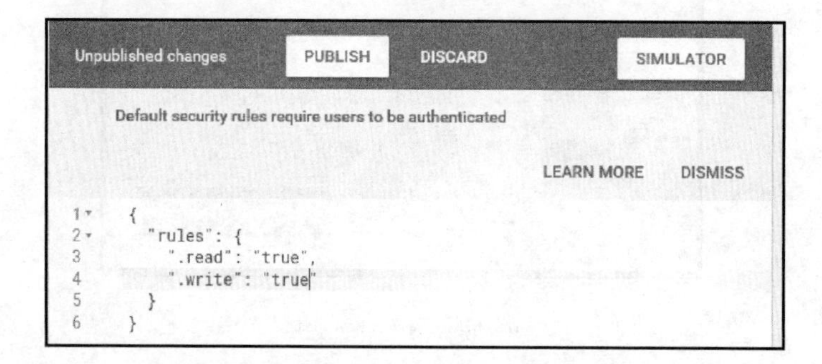

Removing read/write security from database

8. After you make the changes to the values, you will then need to click on the **Publish** button at the top of the panel.

Did you note that everything in the database looks like JSON? That's because the real-time database is entirely in JSON.

9. Navigate back to the **Data** tab. Note the comment about default security rules. Again, we want to disable security for this sample test, so click on the **Dismiss** link beside the notice. This will disable the first level authentication.
10. At the top of the **Data** tab will be the project URL, which should contain the name you set for the project (`testapp`). Copy the entire URL to your clipboard by selecting the text and then pressing *Ctrl* + *C* (*command* + *C* on Mac). Be sure to leave the Firebase console open in your browser window.

Firebase supports standard OAuth, custom OAuth, Google, Facebook, Twitter, and federated authentication support. It also supports multiple levels of database rules you can apply to the node level.

Now, with the Firebase real-time database configured, we will jump back to Unity and get the project configured:

1. Double-click on the `UIHandler` script located in the `Assets/TestApp` folder in the **Project** window to open the script in your editor of choice.
2. Scroll down in the file and locate the `InitializeFirebase` method, as shown in the following code, for review:

```
void InitializeFirebase() {
    FirebaseApp app = FirebaseApp.DefaultInstance;
    app.SetEditorDatabaseUrl("https://YOUR-FIREBASE-
    APP.firebaseio.com/");
```

3. Select the URL text, shown highlighted, and paste the URL you copied earlier to your clipboard by pressing *Ctrl* + *V* (*command* + *V* on Mac). Review the previous set of steps if you need to copy that URL to your clipboard again.
4. Save the file and return to Unity. Wait for the script to recompile and click on Play to run the test app in the editor.

5. An UI will be displayed in the **Game** window. Enter an e-mail address and score then click on the **Enter Score** button. Try this a few times to enter a number of scores. Note how the scores populate in the list, as shown in the following screenshot:

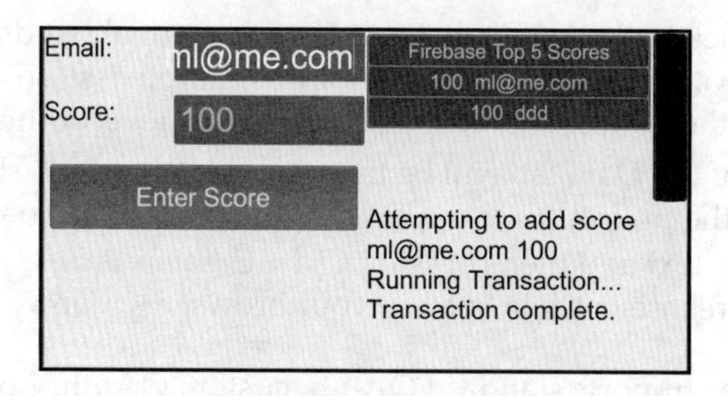

TestApp interface to test adding scores to a real-time database

6. Stop and then start the game again. Note how the scores are added back in automatically. After you are done testing, leave the game running in the editor.

So, it looks like the simple app is saving scores to a database, but now we want to confirm that the data is saved to the cloud database and it updates clients in real time. Follow the given instructions to test the real-time functionality of the database:

1. Go back to the browser window with your Firebase console. The first thing you will note is that there is a new child, called `Leaders`, added to the database. Expand the following nodes and confirm that the entries match the sample data you entered in Unity, as shown in the following screenshot:

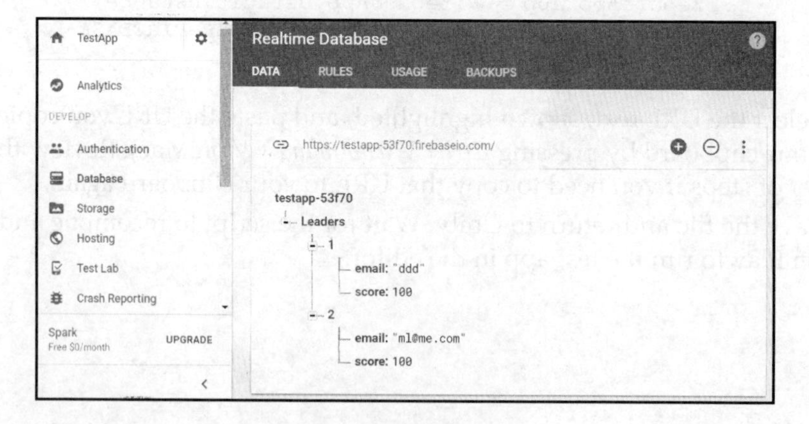

Database Leaders added by Unity

2. Edit the values directly in the **Console** and then switch back to the running Unity session. Note how the values on the score list are changing almost instantly.
3. Try adding or deleting values to the `Leaders` list in the **Console**. Note how they are also automatically updated on the Unity client. This is the power of the real-time database and the automatic client updates.

At this stage, if you have used other network solutions, you may be only somewhat impressed. After all, Photon PUN and UNET support very similar things out of the box and they also provide mechanisms to update transform objects and other Unity components automatically. However, as we previously mentioned and discussed, those automatic transformations are of little value in a location-based game, where the player is always at the center of the game world.

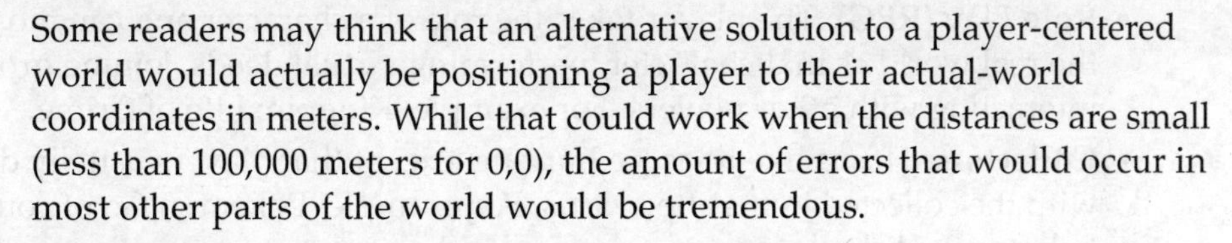

Some readers may think that an alternative solution to a player-centered world would actually be positioning a player to their actual-world coordinates in meters. While that could work when the distances are small (less than 100,000 meters for 0,0), the amount of errors that would occur in most other parts of the world would be tremendous.

Of course, location-based games are not the only place well-suited for a real-time cloud database. Some other suitable genres could be board, empire building, strategy, puzzle, card, and more. Speaking of other types of games, how about we look at suggestions for different types of location-based games in the following section?

Other location-based game ideas

As you may have already guessed, probably in `Chapter 1`, *Getting Started,* Foody GO was a clone of a very popular game franchise that was launched dramatically last year. While the demo game was meant to show you how that popular game was constructed, hopefully it intrigued you into thinking about building your own game. Of course, you wouldn't want to copy a popular game like we did, so let's look at some other options for possible location-based games, as follows:

- **Strategy** (**Empire Building**): Perhaps your players take the role of a king, ruler, or business mogul with the intent of building or securing the resources around them. For example, Parallel Kingdom, Parallel Mafia, Resources, and Turf Wars.
- **Paranormal** (**Survival**): Perhaps you see your players as ghost or zombie hunters that either need to hunt the paranormal or just avoid being eaten. For example, Zombies, Run, and Spec Trek.

- **Treasure Hunting**: Perhaps your players are treasure hunters who need to scavenge for clues and search out areas for hidden treasure, be it real or virtual. For example, Geocaching, Zaploot.
- **Spy (Espionage)**: Your players work for a secret organization and they are sent on missions around the area to discover, explore, and prevent other factions from succeeding. For example, Ingress, CodeRunner.
- **Tower Defense**: Allow your players to protect areas around their neighborhood from outside virtual attacks in a real-world meets tower-defense game. For example, Geoglyph.
- **Hunting (Tracking)**: This is not a game for younger players. Your players play the role of hunter and hunted. The hunters must move within a specific physical distance to catch a hunted player. For example, Shift.
- **Role Play (RPG)**: The player takes the role of a character and can move around the real world as that character, performing virtual deeds, joining groups, and interacting with other players. For example, Kingpin: Life of Crime.
- **Collectors**: This is like treasure hunting, except that players usually do something with the collected items afterwards. For example, Pokemon Go of course, it had to be on the list.

There are some great ideas in the mentioned list and there will surely be other great ideas as the location-based genre evolves. The future of the genre is a great subject for the next section of this chapter.

The future of the genre

With the popularity of Pokemon Go and the cultural shift that occurred because of it, one can only assume that the location-based and AR genres are now mainstream gaming platforms. Previously, GPS functions were only rarely used on mobile devices, and now they have become mainstream.

This shift will not only encourage better GPS capabilities on mobile devices, but that in turn will allow for apps and games to consume more accurate location data. Also, GPS power consumption will likely be reduced and both these factors will have a positive effect for those developing location-based apps or games.

As apps and games begin to consume more GPS and GIS data from services such as Google or others, it could be assumed that prices will be reduced and limitations on data will be increased. That may sound counter-intuitive, but Google is able to supply those services cheaply because of usage tracking. Google will then typically use that usage tracking to support services like traffic flow analysis or other services. For Google, the more users it tracks, the better the data metrics are. This is why Google Maps is completely free for Android or iOS apps.

Considering these factors, it strongly suggests that the location-based genre will only become more popular for gamers and developers. It will be interesting to see what the next big titles in the genre will be.

Summary

This chapter was a wrap up of the development efforts we undertook through most of the book. We first looked at what items would need to be completed in order to ship the Foody Go game. Next, we looked at important skills that would be valuable to spend time learning. After this, we took a good look at the guidelines to acquire Unity assets, which may fill in missing skillsets. Following this is a quick exercise on how to manage and clean up any unwanted assets lingering in your project. From there, we covered some guidelines to release a game. Then, we switched gears and tried to uncover all the problems that a developer faces when building a location-based game; this took us into a discussion of supporting multiple players in our game. As we have seen though, there were a number of unique problems with location-based games that make it improbable to use off-the-shelf solutions. Instead, we looked at the possibility of implementing custom multiplayer solutions and then took a serious look at a real-time cloud-based database from Google, called **Firebase Real-time**. We ran through a quick exercise to set up and configure Firebase for use in Unity as a multiplayer platform. From there, we took a brief look at some other location-based game ideas for inspiration. Finally, we looked at what the future may bring for the location-based genre.

The next chapter is dedicated to troubleshooting problems that you may have encountered while building the demo game. There will also be some good practical knowledge shown about debugging in Unity, console tricks, and other troubleshooting tips.

Summary

10
Troubleshooting

Every developer will, at some point, face an unexpected issue or problem that blocks their progress to further development. It could be something as malignant as a hidden syntax error or far more serious. Either way, the developer must use the tools at their disposal and work past the problem. This chapter will take a look at introducing or familiarizing you with the various troubleshooting tools available to mobile Unity developers. After that, we will take a more advanced look at some specialized tools to use when the problem seems especially difficult. Of course, we will also want to cover options for tracking issues or even preventing them. Then, at the end of this chapter, there will be a table provided for reference, to assist you in solving issues that you may encounter working your way through the chapters in this book. Here is a list of the major topics we will cover in this chapter:

- **Console** window
- Compiler errors and warnings
- Debugging
- Remote debugging
- Advanced debugging
- Logging
- CUDLR
- Unity Analytics
- Issues and solutions by chapter

If you have jumped ahead to this chapter from a previous section of the book, and you can't get past due to an issue, then please jump ahead to the last section in this chapter, *Issues and Solutions by Chapter*.

Console window

The **Console** window should be your starting place whenever you encounter an issue. It can be accessed from the menu by selecting **Window** | **Console**, which will open the **Console** window that you can dock within the editor as you see fit. Depending on your preference and experience, you may want to always have the **Console** visible. Either way, as soon as something goes wrong, it certainly should be the first place you check.

Let's take a quick tour of the **Console** window in some detail, as it will be central to several other elements in this chapter:

- Be sure to have the Unity editor open. If the **Console** window is not open, then open it from the menu by selecting **Window** | **Console**.
- Take a look at the window and familiarize yourself with the buttons and context menu. The following is a screenshot of the **Console** window with the typical configuration:

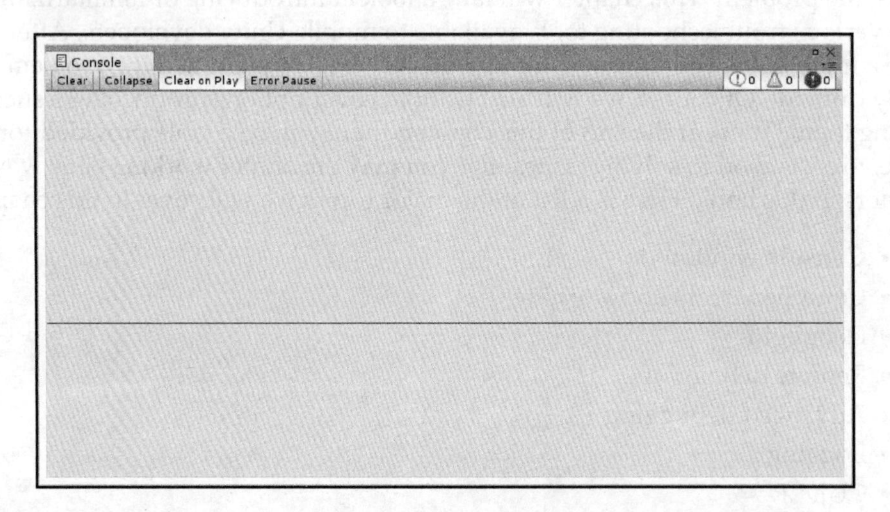

The Console window with typical configuration

How about we take a look at what each of the buttons do and some helpful hints you may not find in the Unity documentation:

- **Clear**: This clears the current window of all messages. It can be useful to clear your log on extended testing sessions.
- **Collapse**: This will collapse all of the same log messages and identify label with a number. This is especially useful if you have a repeating message you want to track, but don't want the window to be overwhelmed with messages.

- **Clear on Play**: This clears the log on every new session run in the editor. It is very useful and something you will want to have on all the time.
- **Error Pause**: This causes the editor to pause when an error is encountered during play. This is another great feature that lets you catch those errors as they happen. Unfortunately, you cannot isolate the type of error you want to track.

The icons on the right-hand side of the window are described as follows:

- **Info Filter**: This is the first button on the right-hand side. This will filter on/off any info or general debug messages sent to the Console. It is a useful way to block out noise.
- **Warning Filter**: This is the button with the yield sign–a filter you can use to turn warnings on/off, again to reduce noise in the window. Avoid the impulse to turn off warnings; they are helpful reminders that you will want to reduce in number.
- **Error Filter**: This is the last button on the right, and is a filter to turn error messages on/off. Generally, it's a good idea to always keep this filter on. However, on occasion, it may be useful to ignore noisy errors that are filling the window in order to track a missing debug message.

The context menu is explained as follows:

1. Click on the **Console** window context menu icon located in the far top-right corner of the window. This will open the context menu, as shown:

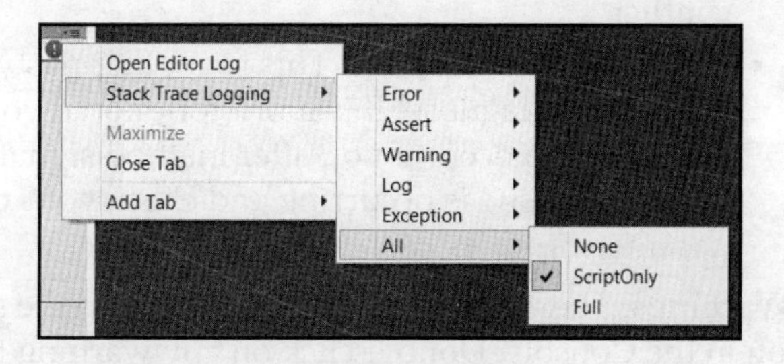

Console window context menu shown

2. In order to be thorough, let's explain the log menu items, as follows:
    - **Open Editor Log**: This will open the detailed log for the editor. We will look at this log in more detail when we get to the section on logging.
    - **Stack Trace Logging**: This sets the amount of stack trace to be included in the Console log. Generally, you will only set this to **ScriptOnly**, as shown in the screen excerpt. Again, we will cover this option in more detail in the sections on debugging and logging.

So, the next time something goes wrong, be sure to start with opening the Console. In the next section, we will look at a set of compiler messages sent to the Console.

Compiler errors and warnings

Compiler messages will fill the **Console** window whenever the project game scripts are recompiled, which may happen as a result of an asset import or script change. The more severe compiler errors will block your ability to run the project in the editor. Warnings are more benign and less critical, but it is equally important to observe when they happen. The following list covers some of the more common errors and warnings you may encounter:

- **Compiler Errors**: These will show up as red text in the status bar or with an error icon in the Console:
    - **Syntax error**: This is likely the most common error you will encounter. Double-click on the issue to open the script to the point of the error in your editor. Simply edit the script to fix the syntax issue.
    - **Missing script**: This is a nasty issue that can occur when importing assets–a script may have moved or there could be a naming conflict. A missing or broken script will be removed as a component from any game object. You will either want to correct the issue by reimporting broken assets or managing the name conflict.
    - **Internal compiler error**: This is another nasty error that can be difficult to diagnose. It is more common if you are using plugins, but may also occur if you alter method signatures. Try to isolate where the issue is occurring and check your use of methods or parameters.
- **Compiler Warnings**: These appear as yellowish text in the status bar, or with a warning icon in the Console. Double-click on any warning to be taken to the offending code in your editor of choice:
    - **Obsolete code**: Unity will flag code that is using properties or methods that have been deprecated. This can be a common issue if you use older assets not released for your version of Unity. You will need to update the code to use the new method in order to remove the warning.

- **Inconsistent line endings**: This is an annoying warning that may occur from switching editors or importing code. Fix the issue in your code editor by setting consistent line endings by navigating to **MonoDevelop**: **Project** | **Solution Options** | **Source Code** | **Code Formatting**, **Visual Studio**: **File** | **Advanced Settings**.
- **General warnings**: This is like unused fields or variables, not critical, but something to clean up when your game or scripts are ready for release.

There is a way to turn compiler warnings into errors and thus force all warnings to be fixed. While this may be a preference for some developers, it is not recommended as the best practice.

A good habit to get into is to clear the **Console** window before script editing or asset importing. That way it is easier to track and filter compiler issues. After you fix those compiler issues, it becomes time to track down any runtime errors or warnings, which we will cover in the next section.

Debugging

Unity provides a great interface in the editor for watching and editing the state of game objects and components while your game is running. While this will likely be the way you can debug most of your game, you may still encounter times where logic in a script needs a finer look. Fortunately, Unity also provides an excellent set of tools to allow you to debug your scripts in the editor while your game is running. Let's look at how you might start a debugging session:

1. Open Unity with an empty project.

If you are using Visual Studio, this exercise assumes that you have already installed the tools extension and configured the editor preferences.

2. From the menu, select **Assets** | **Import Package** | **Custom Package...** and navigate to the `Chapter_10_Assets` folder in the book's downloaded source code. Select `Chapter10.unitypackage` and click on **Open**.
3. The package is small, so it should import quickly. Click on the **Import** button on the **Import Unity Package** dialog to continue.

4. Locate the `Main` scene in the `Assets/Chapter 10/Scenes` folder in the **Project** window and double-click on the scene to open it.
5. Locate the `RotateObject` script in the `Assets/Chapter 10/Scripts` folder in the **Project** window and double-click on the script to open it in your editor of choice.

We will demonstrate the use of the debugger for MonoDevelop and Visual Studio. If you are using another code editor, the process may be different.

6. Set a breakpoint on a single line of code inside the `Update` method, as shown in the following screenshot:

```
RotateObject ▸ Update ()
1  using UnityEngine;
2  using System.Collections;
3
4  public class RotateObject : MonoBehaviour {
5
6      private float angle;
7      // Use this for initialization
8      void Start () {
9
10     }
11
12     // Update is called once per frame
13     void Update () {
14         transform.rotation = Quaternion.AngleAxis (angle++, Vector3.up);
15     }
16 }
17
```

7. The next step will depend on the editor you are using. Follow the directions for your selected editor to start debugging:
    - **MonoDevelop**: In the toolbar, press the Play button to start debugging. The **Attach to Process** dialog will show; now select the Unity process your project is loaded in and click on **Attach**.

If you have multiple Unity editor instances running, it may be a bit of trial and error to find the right process to attach to. Pay special attention to the process ID after you attach to the correct process. Of course, the alternative is to only run a single instance of Unity.

- **Visual Studio**: In the toolbar, press the Play (**Attach to Unity**) button to start debugging. Visual Studio is smart enough to attach itself to the Unity editor on its own.

8. Return to Unity and run the project by pressing the **Play** button.

9. In a very short time, you will be taken to your script editor and the breakpoint you set earlier will be highlighted. At this point, use your mouse to hover over the text and inspect any variable properties, as shown in the following screenshot:

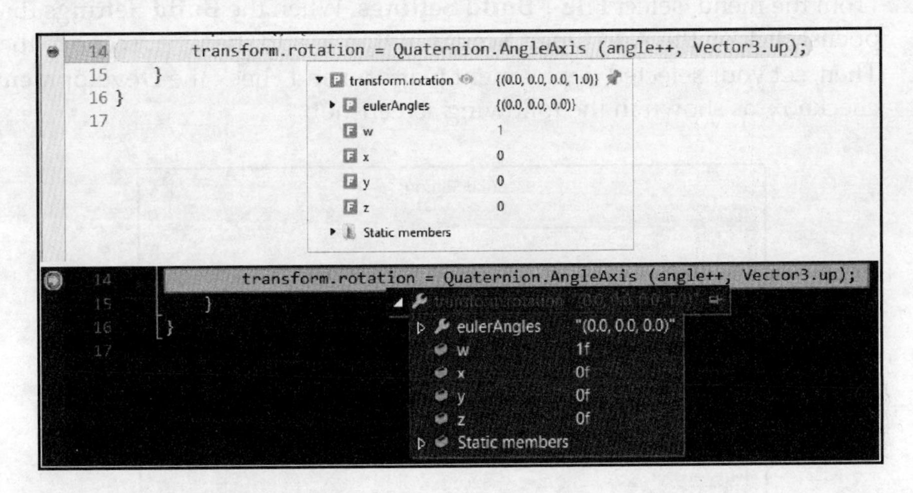

An example of breakpoint and type inspection in MonoDevelop and Visual Studio

There are a number of other debugging options you can employ, such as watches, but the preceding exercise will get you started. Of course, there are times when you are deploying to a mobile platform where you want to debug directly on the device; we will cover that in the next section.

Remote debugging

Being able to debug your project's scripts in the editor is a great feature. Of course, being able to debug your script as it is deployed onto a target platform is like the Holy Grail for developers. No longer are you faced with unknowns about how your script/code is running on a device when you can actually watch the code run with remote debugging.

Remote debugging is a powerful feature that has been around for a while, but it does have limitations and may introduce its own connectivity problems. Before attempting to remote debug an application, be sure that you can deploy it to your mobile device without issues. If you have issues deploying your project to a device, this section will not help you. Instead, refer to the section on *Issues and Solutions by Chapter* at the end of this chapter.

Follow the given instructions to set up remote debugging in your editor:

1. Use the project from the previous debugging exercise to get started. If you jumped here, then just follow steps *1–4* to open the project and set the scene.
2. From the menu, select **File | Build Settings**. When the **Build Settings** dialog opens, click on the **Add Open Scenes** button to add the `Main` scene to the build. Then, set your selected deployment platform and check the **Development Build** checkbox, as shown in the following screenshot:

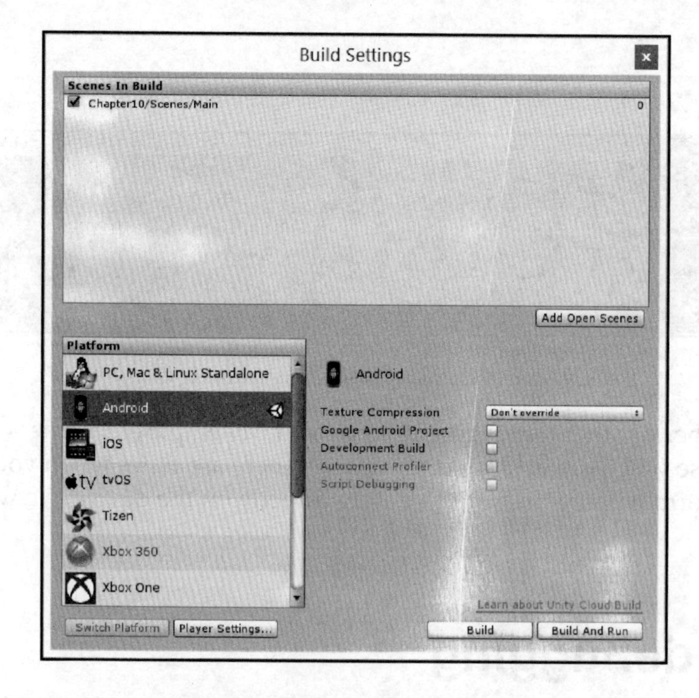

Build Settings dialog with scene added and platform chosen

3. At this point, you may need to set additional **Player Settings** depending on your specific platform. Click on the **Player Settings...** button to open the **PlayerSettings** panel in the **Inspector** window. Then, set the required settings for your chosen platform, iOS or Android.
4. The next step will depend on what platform you are deploying your game to:
    - iOS:
        - Ensure that your mobile device is connected to the same network (Wi-Fi) as your development computer.
        - After you deploy the game, you can disconnect the USB cable.

- Android:
    - Open a Terminal/CMD prompt and navigate to your `Android SDK/platform-tools` folder
    - Enter the following command:

```
adb tcpip 5555
```

- This should output the following message:

```
restarting in TCP mode port: 5555
```

- Open your device and find the IP address by opening **Settings | About | Status**
- Write down or remember the IP address, and then enter the command, replacing your device IP address using following command:

```
adb connect DEVICEIPADDRESS
```

- This should respond with the following message, replaced with your IP address:

```
connected to DEVICEIPADDRESS:5555
```

- Confirm that the device is installed by running the following command:

```
adb devices
```

- This should output something like the following (note that you may see an entry for your device by name):

```
List of devices attached
DEVICEIPADDRESS:5555 device
```

5. Click on the **Build and Run** button on the dialog and save your file according to the name you set in the package identifier.

6. After the simple demo loads and runs on your device, disconnect the USB tether. Then, return to your script editor. Use the instructions for your editor found in the following table:
    - MonoDevelop:
    - Press the Play button or *F5* to start debugging
    - Select the entry that matches your platform and device and click on the **Attach** button

The game on your device will pause when a breakpoint is hit, and you can debug the game as you would locally.

- Visual Studio (2015):
- From the menu, select **Debug | Attach Unity Debugger**
- A **Select Unity Instance** dialog will be displayed and it will list the instances you can connect to, shown as follows:

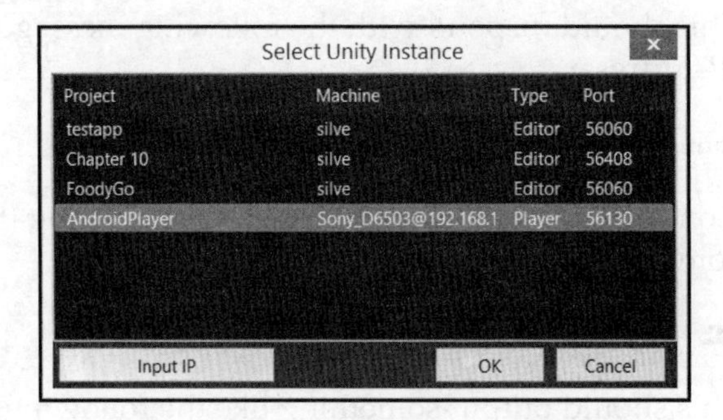

Dialog showing Unity instances

- Select the instance that matches your device (*Type=Player*) and click on **OK**
- You can now use the debugger as you would have locally

7. Note that, while remote debugging, you may need to wait for certain actions, such as variable inspection, so be patient. This happens because the editor needs to retrieve the state from the remote device and not locally.
8. When you are done debugging, click on the Stop button on the MonoDevelop or Visual Studio. This should remove the debugger and let the game continue running. On occasion, if the editor (usually MonoDevelop) gets locked, you have to close the editor.

Remote debugging is a great option for debugging code while developing, but in some situations you may want a few more advanced tools at your disposal. In the next section, we will cover a couple of advanced tools for debugging mobile applications.

Advanced debugging

No matter how much time you spend in the Unity editor, it is always helpful to have some more advanced debugging options, especially those options that can be run completely disconnected from your development machine. The following is a table of some advanced debugging tools you may want to consider using:

Tool	Ease of Use	Description	Source/Link
HUDDebug	Easy: download from the asset store	This creates a great integrated debugging tool on the platform. It supports a console, FPS, memory, heap, and extensions.	Search for HUDDebug on the Unity Asset store
Unity Remote 5	Easy: Difficult – may experience connection issues	This is a fantastic tool when it works. It will allow you to run a game on a mobile device, and track UI and game input with the editor. However, it can be problematic to run because of connection issues. Hopefully, this tool will overcome these issues in future releases.	Search for Unity Remote in your App store
Charles Proxy	Medium: Difficult	Charles Proxy allows you to configure the network traffic from your mobile device to route through your development machine and monitor it. If you have issues with your mobile device making calls to web services or you are doing advanced networking, this is the tool for you. While the tool is not free, it will be essential to understanding networking issues.	`www.charlesproxy.com` Search for Charles Proxy setup on your preferred device

While debugging is certainly a tool you can use to resolve issues, it isn't something you want to do all the time. The better approach to understanding how your game is running is to add logging, which we will cover in the next section.

Logging

If you have already covered a few of the chapters in this book, you can appreciate how valuable logging is to making sure your game works as expected. In Unity, all log messages are output to the Console unless you create a custom logger that writes to a file or service. We will create a custom logger later in this section. For now though, let's look at the logging options Unity provides us with out of the box, as listed:

- `print`: This is the shorthand equivalent to `Debug.log`.
- `Debug.Log`, `Debug.LogFormat`: This outputs the standard information message in unformatted or formatted text. Messages appear with info icon in the **Console** window.

- `Debug.LogError`, `Debug.LogErrorFormat`: This outputs unformatted and formatted error messages. Messages appear with an error icon in the **Console** window.
- `Debug.LogException`: This outputs the exception to the console with an error icon.
- `Debug.LogWarning`, `Debug.LogWarningFormat` : This outputs unformatted and formatted warning messages. Messages appear with a warning icon in the **Console** window.
- `Debug.LogAssertion`, `Debug.LogFormatAssertion`: This outputs unformatted and formatted test assertion messages.

In order to use these logging capabilities, you would add these statements to entry, exit, or other points in a script. The following is an example of Unity script that shows how each of these logging types may be used:

```
using UnityEngine;
using System.Collections;
using System;

public class LoggingExample : MonoBehaviour {

    public GameObject target;
    public float iterations = 1000;
    private float start;
  // Use this for initialization
  void Start () {
        Debug.Log("Start");

        if (target == null)
        {
            Debug.LogWarning("target object not set");
        }
        if (iterations < 1)
        {
            Debug.LogWarningFormat("interations: {0} < 1", iterations);
        }

        Debug.LogFormat("{0} iterations set", iterations);
        start = iterations;
  }
  // Update is called once per frame
  void Update () {
        //try/catch used for demo
        //never use in an update method
```

```
        try
        {
            iterations--;
            Debug.LogFormat("Progress {0}%", (100 - iterations / start *
    100));
        }
        catch (Exception ex)
        {
            Debug.LogError("Error encountered " + ex.Message);
            Debug.LogErrorFormat("Error at {0} iterations, msg = {1}",
    iterations, ex.Message);
            Debug.LogException(ex);
        }
      }
    }
```

The `LoggingExample` class is an example of the use of the various types of logging available in Unity.

In the `LoggingExample` class, what will happen if the initial `iterations` is set to 0? What change can you make to get the sample to throw an exception?

Of course, most of the time, outputting log messages to the **Console** window will be fine, especially in development. However, in other situations, after the game has been deployed, either for testing or commercially, you may still want to track those messages. Fortunately, there is a very easy ability to handle custom log output, as shown in the following class:

```
using System;
using System.IO;
using UnityEngine;

public class CustomLogHandler : MonoBehaviour
{
    public string logFile = "log.txt";
    private string rootDirectory = @"Assets/StreamingAssets";
    private string filepath;
    void Awake()
    {
        Application.logMessageReceived += Application_logMessageReceived;

#if UNITY_EDITOR
        filepath = string.Format(rootDirectory + @"/{0}", logFile);
        if(Directory.Exists(rootDirectory)==false)
        {
            Directory.CreateDirectory(rootDirectory);
        }
#else
```

```
        // check if file exists in Application.persistentDataPath
        filepath = string.Format("{0}/{1}", Application.persistentDataPath,
logFile);
#endif
    }

    private void Application_logMessageReceived(string condition, string
stackTrace, LogType type)
    {
        var level = type.ToString();
        var time = DateTime.Now.ToShortTimeString();
        var newLine = Environment.NewLine;

        var log = string.Format("{0}:[{1}]:{2}{3}"
            , level,time, condition, newLine);

        try
        {
            File.AppendAllText(filepath, log);
        }
        catch (Exception ex)
        {
            var msg = ex.Message;
        }
    }
}
```

The `CustomLogHandler` works by attaching itself to the `Application.logMessageReceived` event in the `Awake` method. This event is called whenever content is logged in Unity. The rest of the class is about configuring the correct file path, creating folders if needed, and then formatting output, where the actual logging takes place in the `Application_logMessageReceived` method. While this class may not be well suited for mobile platforms, it is perfect for tracking log running messages in the editor or on a desktop platform deployment. In the next couple of sections, we will look at how this log handling can be used for debugging and/or to release deployments.

CUDLR

If you have already covered a couple of chapters in this book, you may be familiar with CUDLR, which is an excellent remote logging, debugging, and inspection tool. For those of you that missed the CUDLR setup in `Chapter 2`, *Mapping the Player's Location*, no worries, as we will do a review of the setup here. Of course, if you are here because of issues with CUDLR, refer to the final section in this chapter, *Issues and Solutions by Chapter*.

CUDLR is a remote console that runs through an internal web server created within your game. It uses the same technique we used to output the log output, but it also provides for object inspection and even customization. Use the following instructions to set up CUDLR (if you have already done this, you may want to just skim over this section):

1. If you have not already set up the `Chapter 10` project, then open a new Unity project and import `Chapter10.unitypackage` from the `Chapter_10_Assets` folder of the books downloaded source code.
2. Open the **Main** scene from the `Assets/Chapter 10/Scenes` folder.

CUDLR has been included as a part of the `Chapter10` package.

3. From the menu, select **GameObject** | **Create Empty**. Rename the object `CUDLR` and reset its transform to zero.
4. Drag the `Server` script from the `Assets/CUDLR/Scripts` folder in the **Project** window onto the new `CUDLR` GameObject.
5. By default, `CUDLR` sets itself to run on port `55055`. This setting can be changed by inspecting the `CUDLR` GameObject in the **Inspector** window. For now though, leave it at the default value.
6. Press Play in the editor to run the scene. Keep the scene running.
7. Open the browser of your choice and enter the following URL: `http://localhost:550555`

8. The CUDLR window will open in the browser, as shown in the following screenshot:

CUDLR window in a browser (Chrome)

9. At the bottom of the window, enter the following command:

```
help
```

10. This will provide the following output:

```
Commands:
object list : lists all the game objects in the scene
object print : lists properties of the object
clear : clears console output
help : prints commands
```

11. Enter the following command:

```
object list
```

12. This gives the following output:

```
CUDLR
Directional Light
```

```
Cube
Main Camera
```

13. Enter the following command:

```
object print Cube
```

14. This gives the following output:

```
Game Object : Cube
  Component : UnityEngine.Transform
  Component : UnityEngine.MeshFilter
  Component : UnityEngine.BoxCollider
  Component : UnityEngine.MeshRenderer
  Component : RotateObject
```

As we have shown in several sections of this book, CUDLR is also useful for capturing logging activity on a game running on a mobile device. With CUDLR, you could simultaneously track logging output on several devices at the same time. Of course, CUDLR is really only intended for capturing logs while debugging or testing. The internal web server CUDLR creates is not something we would want to ship with our game for numerous reasons.

What if we want to track critical error or exception logs after our game is released? Fortunately, there are a number of options available to do that in Unity and we will look at one of these options in the next section.

Unity Analytics

Out of the box, Unity Analytics is something you should have on anytime you release a game. It is essential to provide feedback about your players, the distribution of your game, and many other metrics. We will cover some of the basics of Unity Analytics in this section and then take a quick deep dive into using the tool to track critical errors and exceptions. Being able to access this information will provide you with better support for your game during off-site testing or release.

Follow the given instructions to enable Unity Analytics on your project:

1. From the menu, select **Window** | **Services**. The **Services** window will open, typically over the **Inspector** window.

Performance Reporting will also report on errors, but that requires an upgraded Unity account.

2. Locate the **Analytics** group in the list and set the toggle button on the right to the **On** position, as shown in the following screenshot:

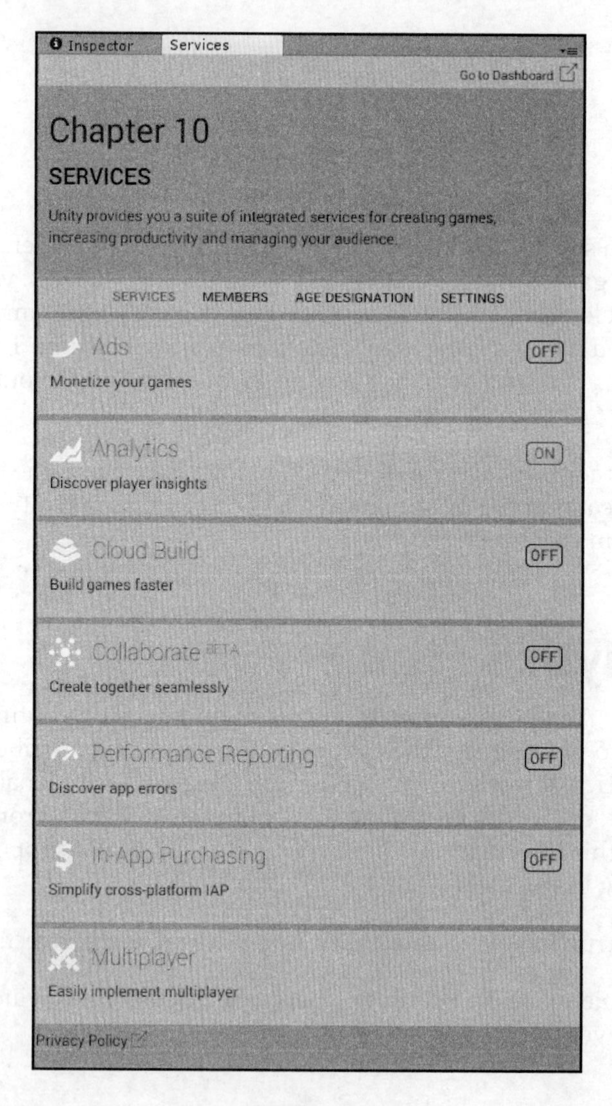

Turning Unity Analytics on

3. Click on the **Analytics** panel in the **Services** window. You may get prompted to confirm the age guidelines of your game. If you do, just select an age greater than 13 and continue.
4. The **Analytics** page will load in the **Services** window. Click on the **Go to Dashboard** button just below the main text. This will open your default browser and take you to the Unity Analytics site. You may have to log in with your Unity account, just do so when prompted.
5. Once the page loads, you should see something close to the following screenshot:

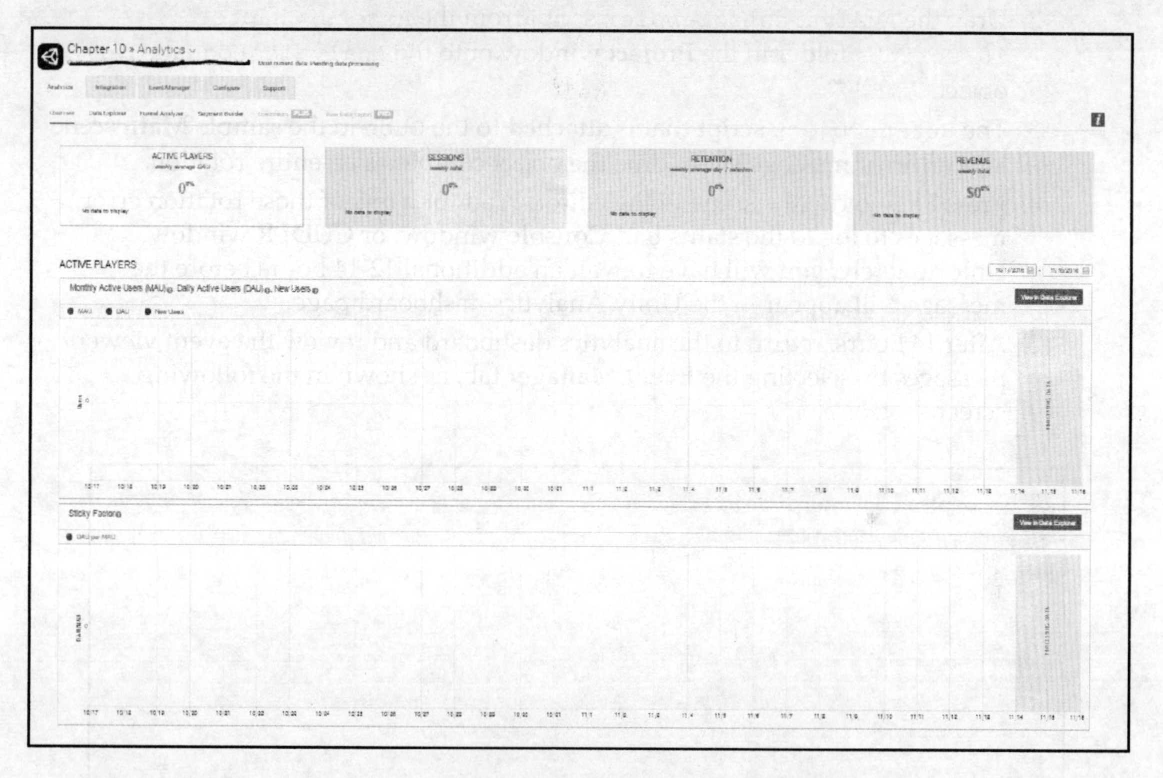

The Unity Analytics project page

Right now, your players, sessions, and other metrics will likely be sitting at 0. This is expected since you likely just enabled analytics on your project. Now, by itself, Unity Analytics won't track logging messages, but we can use some of the custom event mechanisms to track errors or exception logs. Fortunately, all the work for this has been encapsulated into a script called `AnalyticsLogHandler`.

Unity Analytics does not run in real time, which means that you will typically have to wait 12-14 hours to see results. While this makes it essentially useless for debugging, it can be a useful metric when deployed to players globally.

Follow the given directions to set up this script:

1. From the menu, select **GameObject** | **Create Empty**. Rename the object to `AnalyticsLogHandler` and reset the transform to zero.
2. Drag the `AnalyticsLogHandler` script from the `Assets/Chapter 10/Scripts` folder in the **Project** window onto the `AnalyticsLogHandler` object.
3. The `RotateObject` script that is attached to the `Cube` in the sample **Main** scene logs an error message every time the object completes an entire rotation.
4. Press Play to run the scene in the editor. Wait for a few of those rotation error messages to log to the status bar, **Console** window, or **CUDLR** window. Unfortunately, you will have to wait an additional 12-14 hours before the message will appear in the Unity Analytics dashboard page.
5. After 14 hours, return to the analytics dashboard and review the event viewer messages by selecting the **Event Manager** tab, as shown in the following screenshot:

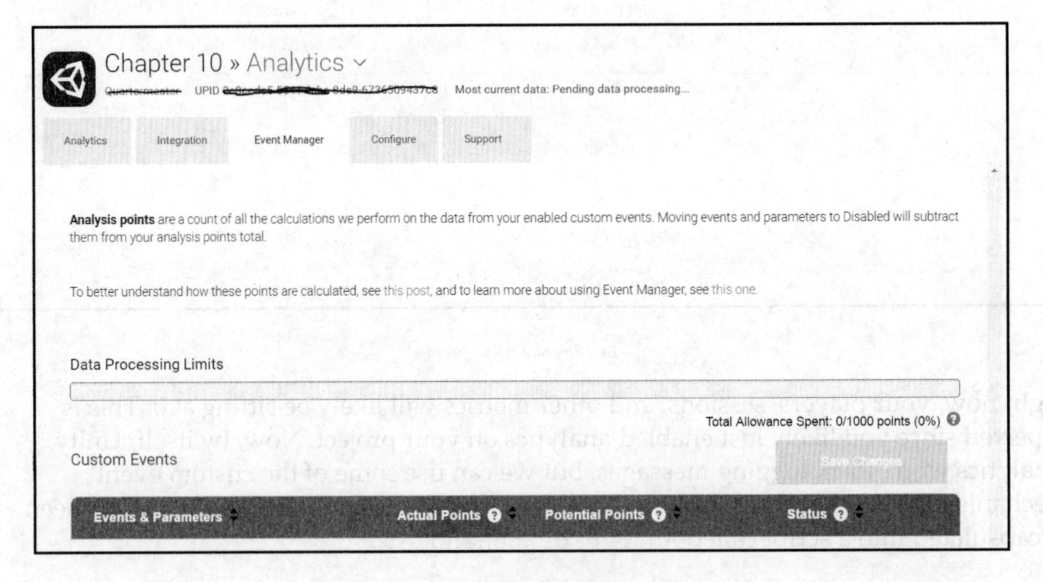

The Event Manager tab on the Unity Analytics dashboard

Let's take a look at the AnalyticsLogHandler script, as illustrated:

```
using UnityEngine;
using UnityEngine.Analytics;

public class AnlyticsLogHandler : MonoBehaviour
{
    public LogType logLevel = LogType.Error;
    void Awake()
    {
        Application.logMessageReceived += Application_logMessageReceived;
    }

    private void Application_logMessageReceived(string condition, string
stackTrace, LogType type)
    {
        if (type == logLevel)
        {
            Analytics.CustomEvent("LOG", new Dictionary<string, object>
            {
                { "msg", condition },
                { "type", type.ToString() }
            });
        }
    }
}
```

This script is simpler, but the implementation is similar to the CustomLogHandler script we looked at earlier. There is one major difference and that is the use of the Analytics object to create a CustomEvent that then automatically gets sent to Unity Analytics. In this case, we are using the method to track logging messages or other error conditions. Of course, you could add that snippet of code anywhere you wanted to track a custom event in your game.

The **Event Manager** uses a system of points to determine how many events you can track and perform other analytics on. Every project starts with 1,000 points, which should cover most of your tracking needs.

Once you start collecting custom events, be it from errors or other regular game activity, you can plot these metrics against other metrics. For instance, you could plot the number of errors per user or as per the number of game sessions, which would tell you how critical or benign an error is and how quickly you may need to respond. Setting up those plots is outside the scope of this book but worth some investigation on the Unity site when you have time to do so.

As you saw, analytics can be a great tool for not only tracking player activity, but also other activities, such as errors or other custom events. In the next and final section of the book, we will look at other issues you may have encountered during the course of this book, with possible solutions, of course.

Issues and solutions by chapter

The following table lists, by chapter, the potential errors you may come across and a list of possible solutions for each issue:

Chapter/Section	Issue	Solution
1. Setting up Android for development	Unable to locate device with the `adb` command	• Ensure that the device is connected by USB • Ensure that the cable is working with other devices • Ensure that the USB port is working correctly; try a different USB device, such as a thumb drive • Confirm that the device has USB debugging enabled • Confirm that the device driver is installed • Unplug and plug in the USB cable a few times, waiting several seconds between connects
1. Building and deploying the game	Project won't build (Android)	• Confirm that the SDK and JDK paths are set correctly • Ensure that the bundle identifier matches the name of the build apk • Confirm that you installed the correct SDK platform; if not, open Eclipse and install other platform versions
1…8. Build and deploying the game	The build gets stuck halfway and freezes	• This is a common issue and is most often noticed after you plug in a device; just be patient or try canceling the build • If this continues to happen, close and reopen Unity

2. Setting up CUDLR	Unable to connect to CUDLR	• Try changing the port CUDLR uses from 55055 to something else (1024 – 65535) • Confirm that the device IP address is correct • Confirm that your URL syntax is correct — `http://IPADDRESS:PORT` • Confirm that your computer or device is not running a firewall that may be blocking the connection; if so, add an exception for the port as required • Ensure that the game is running on the device • Run the game in the Unity editor and try connecting by localhost — `http://localhost:55055`
2. Creating the map tile and/or 2. Setting up the GPS service	Map tiles are rendering question mark images	• Ensure that the latitude/longitude coordinates are entered correctly • If on a mobile device, confirm that location services are enabled • If on a mobile device, confirm that the GPS service is working by installing a test app or using Google Maps • Check whether the URLs getting sent for tile requests are returning images by copying the URL from the Console or CUDLR window to a browser window • Wait for several hours; you could have exceeded the IP usage limits
6. The database	Database does not work when deployed to a device	• Confirm that the plugin settings are correct for your target platform • Ensure that the database version is of the form #.#.# and by default 1.0.0 • If you are deploying to iOS, confirm that you are using IL2CPP • Stop the game, uninstall it from your device, and redeploy the game

7. Setting up the Google Places API service	Places are not shown on map or do not match the location	• Confirm that your GPS service is running correctly and simulation mode is disabled/enabled, as required • Confirm that location is enabled on your device
8. Updating the database	Unable to sell monsters to a location	• Stop the Unity editor and delete the database file from the `Assets/StreamingAssets` folder • Use a SQLite tool to directly edit or verify the database contents. A great tool is DB Browser for SQLite from `http://sqlitebrowser.org/` • Uninstall the game and reinstall, this will reset the database • Catch some lower-level monsters
	Deploying the game to a mobile device	Consult the `Chapter 1`, *Getting Started* sections

Summary

This chapter was all about fixing issues and resolving problems you may encounter in this book or during typical development. We started by looking at the first place you should always look when anything goes wrong in Unity, the Console. From there, we jumped to looking at some typical errors and warnings that may stop you cold in development. This took us into debugging and then remote debugging scripts from your code editor. This was followed by a further look at the logging capabilities in Unity, where we finished with an example of a custom log handler. This took us back to reviewing CUDLR as our local/remote console that we can connect to any platform and track log messages, or even inspect objects. Having spent so much time reviewing logging-development tools, we also looked at using Unity Analytics to capture error/exception messages when the game is released. Finally, we reviewed a list of potential issues and solutions that you may encounter while building the demo game in this book.

This ends our journey into location-based AR game development. Hopefully, you gained insights into numerous areas of game development with Unity. From here, you are encouraged to explore development in augmented reality, GIS, and other advanced features of Unity development. Furthermore, it is hoped that you take the plunge and develop your own location-based AR game.

Index

W